Pitt Series in Policy

and Institutional Studies

SITE UNSEEN

THE POLITICS OF SITING A
NUCLEAR WASTE REPOSITORY

GERALD JACOB

UNIVERSITY OF PITTSBURGH PRESS

Published by the University of Pittsburgh Press, Pittsburgh Pa. 15260
Copyright © 1990, University of Pittsburgh Press
All rights reserved
Baker & Taylor International, London
Manufactured in the United States of America

Library of Congress Cataloging-in-Publication Data

Jacob, Gerald, 1952-
    Site unseen : the politics of siting a nuclear waste repository/
Gerald Jacob.
        p.   cm.— (Pitt series in policy and institutional studies)
    Includes bibliographical references (p.   )
    ISBN 0-8229-3640-2
    1. Environmental policy—Nevada—Yucca Mountain. 2. Radio-
active waste sites—Government policy—Nevada—Yucca
Mountain. I. Title. II. Series.
HC107.N33E553   1990
363.7'289'0979334—dc20   89-40582
                                                              CIP

CONTENTS

ABBREVIATIONS

AEC	U.S. Atomic Energy Commission
AFRs	temporary, away-from-reactor storage facilities
AMFM	alternative means of financing and management
DOE	U.S. Department of Energy
DOT	U.S. Department of Transportation
EA	environmental assessment
EIS	environmental impact statement
EPA	U.S. Environmental Protection Agency
ERDA	U.S. Energy Research and Development Administration
GAO	U.S. General Accounting Office
MRS	monitored, retrievable storage facility
NAS	National Academy of Sciences
NEPA	National Environmental Policy Act
NGA	National Governors Association
NRC	U.S. Nuclear Regulatory Commission
NRDC	Natural Resources Defense Council
NWPA	Nuclear Waste Policy Act (1982)
OCRWM	U.S. DOE Office of Civilian Radioactive Waste Management
OTA	U.S. Office of Technology Assessment
PUC	Public Utility Commission
R&D	research and development
USGS	U.S. Geological Survey
WIPP	Waste Isolation Pilot Plant

ACKNOWLEDGMENTS

▼

In the course of researching and writing this book I received assistance from several faculty members at the University of Colorado-Boulder, including David Greenland (geography), Paul Wehr (sociology), and Susan Clarke (political science) whose insights into political theory and policy making were indispensable. The University of Colorado-Boulder provided me with funding and other assistance for which I am grateful. A special thanks is sent to Andrew Kirby and Sallie Marston, now on the faculty at the University of Arizona, for their encouragement, guidance, and willingness to make their home mine for a while. The staff of Utah's High-Level Nuclear Waste Office—now scattered throughout the West—gave me the opportunity to plunge into the on-camera politics of repository siting and to collect much of the information used in this study. I thank the staff at the University of Pittsburgh Press for all their efforts, with a special thanks to Kathy McLaughlin for her careful attention to the details of this project. Several anonymous reviewers, along with Fred Shelley, provided helpful comments on the manuscript. Kimberly Wille drafted several of the figures. A personal thanks to Catherine Schieve for the flowers. There is no way to adequately thank Kathryn Mutz for years of patience, support, and understanding other than to trust and welcome a future together.

INTRODUCTION

▼

On January 3, 1961, at the National Reactor Testing Station in Idaho a prototype military reactor was temporarily shut down. That part of Idaho, near the easily missed towns of Arco and Atomic City, where Idaho Falls is "city," is a remote stretch even by Idaho standards. As the story goes, what took place out there may have been a deliberate suicide-murder, committed by a reactor operator who suspected his wife of having an affair with another operator.

Three men were working on the down reactor, manually reengaging its control rods, when the reactor went critical. There was a steam explosion. Two operators were killed instantly. A control rod, blown out of the pile, speared one man through the groin and pinned his body to the roof. Those parts of the corpses which had been covered by clothing were buried in lead coffins placed inside of lead-lined vaults. But parts of the bodies directly exposed to radioactive materials—the heads and hands—had to be amputated and stored, awaiting a final burial ground for the nation's high-level radioactive waste.[1]

Decades after this accident, the United States still cannot decide where to bury its most dangerous forms of garbage—the spent fuel rods from nuclear reactors and other radioactive materials which remain toxic for thousands of years. Today the disposal of high-level radioactive waste appears to be an intractable problem of almost incomprehensible proportions. Millions of gallons of liquid wastes continue to be stored in underground tank farms along with millions of cubic feet of radioactive sludge. Thousands of spent fuel rod assemblies fill pools at nuclear power plants. Decades of bomb building have left a sprawling complex of weapons facilities which temporarily store high-level waste. The stuff is found in any state with a commercial or military reactor. This book does not even address the millions of tons of irradiated clothing, tools, medical instruments, and building materials classified as "low-level" waste or "mid-level" transuranic wastes which include plutonium-contaminated materials. The United States is only beginning to confront what to do with the mammoth bulk of waste produced by decommissioning commercial reactors and weapons facilities.[2]

In the early years of nuclear power, optimism overshadowed any concern that the disposal of radioactive waste would present a serious problem. Scientists, technicians, and administrators were given a free hand to speed the commercialization of nuclear reactors. By the late 1970s, however, uncertainties about the disposition of spent reactor fuel, combined with questions about the safety of nuclear reactors, threatened the future of nuclear power and its promoters. State and local attempts to regulate nuclear facilities kindled conflicts over local authority to halt the construction of new nuclear power plants. Stories in the mass media about reactor accidents, nuclear waste mismanagement, cost overruns, and the alleged federal cover-up of radiological disasters created public distrust which threatened the continued promotion of nuclear power by the federal government. The boundary between commercial and military nuclear programs was never clear, but proposals to recover plutonium from commercial wastes to use in the production of nuclear weapons only further obscured it. Demands for greater public access to bureaucratic decision making and government documents were transformed into demands for closer oversight of nuclear programs. Responsibilities of the once-powerful Joint Committee on Atomic Energy were distributed among congressional committees on energy, environment, and interior whose members were more critical of nuclear industries. Public access to federal nuclear agencies increased with the division of the Atomic Energy Commission (AEC) into the Nuclear Regulatory

Commission (NRC) and the Energy Research and Development Administration (ERDA)—later reconstituted as the U.S. Department of Energy (DOE). By 1980 political-economic relationships, which heretofore had favored the development of nuclear energy, were faltering.

High-level waste became a major topic in the debate over the future of nuclear power—a debate heard in state legislatures, the mass media, and Congress. Utilities threatened power plant shutdowns and blackouts in metropolitan areas unless the federal government made good on a promise to build a storage facility for spent fuel rods from commercial reactors. Some feared that research, nuclear medicine, and cancer treatments would end unless a disposal site was found. Idaho demanded that the federal government make good on its promise to remove waste from the Rocky Flats weapons plant near Denver that was temporarily stored at the Idaho National Engineering Laboratory. New York had liquid waste from an abandoned fuel reprocessing plant in need of disposal. Added to this were concerns that the by-products of the peaceful uses of atomic energy could be used to fuel the proliferation of nuclear weapons.

Engineers assured the country that the technology for safely sealing spent fuel and other high-level waste in deep mines, or repositories, was in hand (fig. 1). In their view, only lack of political will stood in the way of a solution. In 1976 testimony before congressional committees and articles in professional journals, Union Carbide—a DOE contractor—identified the Nevada Test Site, Utah's Paradox Basin, the Texas panhandle, and the Gulf Coast salt domes as likely sites for the nation's first high-level nuclear waste repository.[3] Shortly after passage of the 1982 Nuclear Waste Policy Act (NWPA), the DOE identified nine sites as potentially acceptable for the first high-level waste repository (fig. 2). Following the production of thousands of technical reports, DOE released its draft environmental assessment of potential repository sites. Two years later, after receiving thousands of public comments on the draft, it issued the final environmental assessment. Sites in Nevada, Washing-

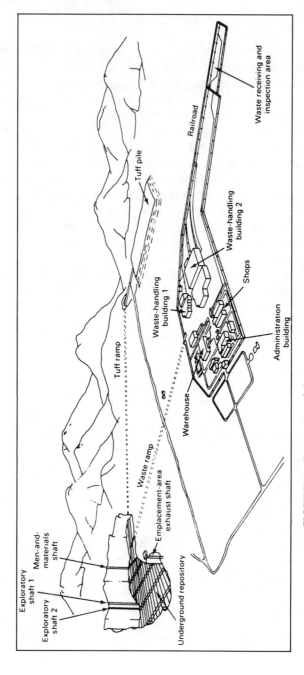

FIGURE 1. Perspective of the proposed repository at Yucca Mountain, Nevada.

ton, and Texas were chosen for additional, technical studies called "site characterization." Despite years of research and debate it appeared that little had changed. The DOE chose the same sites it had identified years before the site selection criteria required by the NWPA were even drafted. In addition, these same sites had been repeatedly identified as potential sites throughout years of congressional debate and testimony predating the NWPA.

In 1987 Congress revisited the NWPA, amended it, and restricted site characterization activities to Yucca Mountain, Nevada (adjacent to the Nevada Test Site). This effectively left Nevada with the site of the nation's high-level waste repository unless some fatal flaw was discovered in the course of site characterization. In retrospect, the choice of Yucca Mountain for the nation's first high-level nuclear waste repository appears to have been neither systematic nor the result of an organized program to identify the best site. During the years of hearings and debate on a Nuclear Waste Policy Act, the states that would host deep geologic repositories were already known. There were few doubts about the outcome of the post-NWPA site selection process. States such as Ohio, New York, and Michigan held potentially suitable geologic formations but stood little chance of being selected. Opposition to the DOE's repository program was in progress in some states that were possibilities for site selection long before passage of the 1982 NWPA. The fight over where to site a high-level waste repository was actually part of a much larger conflict extending far beyond the issue of site selection criteria. The conflict continues today following amendments to the 1982 Nuclear Waste Policy Act and the choice of the site in Nevada.[4]

Overview

This study of the siting of a nuclear waste repository converged on an interpretation of the conflict as one between those who would reassert the authority, organization, and priorities of "the nuclear establishment," and those who de-

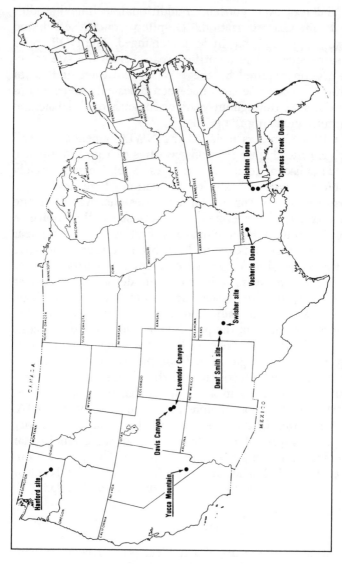

FIGURE 2. Potentially acceptable sites for the first repository. Underlined sites were those selected for characterization.

manded the creation of new patterns of organization, inter-action, participation, and power. The conflict emerged from historical changes which threw past relationships among government, industry, and civil society into disarray. The formation and implementation of the 1982 Nuclear Waste Policy Act was an attempt to restore order to those relation-ships. In this way, siting of a repository is viewed as one part, albeit a key one, of an attempt to restore the authority, credibility, and financial health of those who promoted and benefited from the production of nuclear power.

With this in mind, three major issues form the core of the conflict:

1. *Credibility.* Defined as access to, control of, and confidence in the expertise used in attempts to establish the scientific and technological credibility of the solution.
2. *Legitimacy.* Defined as acceptance of the political process used to define the problem, identify a solution, and imple-ment that solution.
3. *Financial security.* The cost of environment quality and pub-lic health versus the priorities of industry and the economy.

By now it should be apparent that this book is not a legislative history of the 1982 NWPA; nor is it an explana-tion of the geotechnical and environmental questions raised by the construction of a repository at a given site.[5] Rather than discussing questions such as the merits of building a repository in Nevada, the interest here is in larger questions. Why did nuclear waste management become an issue and topic of debate in the first place? How, over time, did the nature of American politics affect the definition of the prob-lem and the solution that emerged? Answers to these ques-tions are found in the context within which the United States approached the management of high-level radioactive waste. The aim here is to account for the appearance of the problem; the process whereby a feasible solution becomes defined in technical and locational terms; the aim for a na-tional, legislated solution; and the structure of programs for implementing that solution.

Even though the conflict over nuclear waste disposal has gone on for years, the choice of a repository site in Nevada appears to have been ordained years ago. Was the outcome ever in doubt? The evidence suggests the answer must be "rarely." One is pushed to the conclusion that, in the course of policy making, known sites and technologies were defended, threats to the existing political-economic relations contained, and the continuity of existing administrative programs preserved. In the end, substantial innovation—either technical or institutional—was impossible. A long-established coalition of industrial, political, and economic interests opposed the construction of new management agencies and successfully kept such proposals off the legislative agenda. Narrowing the policy debate to questions of technology and geography ultimately reasserted the power of that coalition. The steps leading to this conclusion are outlined in greater detail in the chapter-by-chapter outline that follows.

Chapter 1 introduces the coalition of interests described as the "nuclear establishment." More than the U.S. Department of Energy, the nuclear establishment crosses the boundaries of the state, economy, and civil society. The nuclear establishment is not a completely unified political organization but a conglomeration of interests and organizations. Rather than characterizing it as a united front, it should be viewed as a relatively stable pattern of relationships among organizations that supported the production of nuclear power and the promotion of nuclear technologies, products, and policies. This chapter also explains how spatial, historical, and social-organizational themes are merged to develop a perspective on the politics of nuclear waste and a portrait of the context within which nuclear waste policy evolved.

Chapter 2 includes a description of the members and network of relations which constitute the functioning nuclear establishment.

The crises and disorder which produced demands for congressional intervention and national nuclear waste legislation are explored in chapter 3. Consideration is given to how actions by state and local governments and demands for changes in the decision-making process produced a crisis in nuclear waste management. At the same time, changes in the structure of the electric utility industry threatened the economics of nuclear power. These conditions, along with public challenges of establishment expertise and administrative programs, disrupted the functioning of the nuclear industries. Against this background of disorder is superimposed the central conflict and theme of this analysis—the tension between conservative forces of reassertion and the disruptive forces of reconstruction.

In chapter 4 various proposals for reasserting the legitimacy, credibility, and economic health of institutions within the nuclear establishment are contrasted with proposals for constructing new political-economic relations and the institutions responsible for nuclear waste management.

In chapter 5 we consider the 1982 Nuclear Waste Policy Act in light of these contradictory proposals by discussing which recommendations for reform and reassertion were ultimately adopted, which were discarded, and which received no admittance to legislative agendas. On balance, the NWPA deferred to the expertise, authority, and financial requirements of members of the nuclear establishment. Transformation of administrative or economic institutions was avoided. The NWPA confirmed the past decisions, investments, expertise, and authority of the nuclear establishment. The DOE and its relations with other state and economic institutions are also discussed. Radical changes would have threatened the complex of institutions tied to DOE plans for emplacement of spent fuel in a geologic repository. Such changes would also have threatened states with nuclear reactors that supported a federal program to remove dangerous spent fuel rods and other waste from their backyards. Protection of nuclear utilities' investments required the continuation of federal nuclear waste programs. Nuclear waste disposal was associated with dependable energy supplies and thus with national strategic power and regional economic growth. Huge capital investments in power plants, uranium enrichment and fuel fabrication plants, and other public and private facilities weighed in favor of a Nuclear Waste Policy Act that maintained existing institutional arrangements and past commitments to repository sites and technologies. In short, no one began an examination of high-level nuclear waste sites or management with a clean slate.

Reassertion of control over nuclear waste disposal took two forms—the site selection process and the reliance on existing waste disposal technologies. Chapter 6 explains how a federal monopoly of nuclear research and technology influenced political debate by focusing attention on available technological solutions. Centering on past technological investments, and the economics of a quick solution, limited policy options and set an agenda which reinforced old, and excluded new, problem definitions and solutions. Once a federally managed geologic repository was confirmed as the only feasible technical solution, the role of the government expert was reasserted. The goal of removing barriers to the rapid implementation of this technology resulted in programs which reasserted the control and authority of existing institutions and expertise.

In chapter 7 this argument is developed further in an analysis of how commitment to an early siting decision recast the issues. By focusing on site-specific issues, the targets of political opposition were fragmented. Individuals and groups who demanded new social priorities, new institutions, and a redistribution of political power were undermined in their position. These arguments no longer seemed relevant to the site-specific topics crowding the agenda. Finally, narrowing the political agenda to exclude national issues reduced the potential number of recognized interests and participants.

As the post-1982 DOE program encountered increasing opposition, a deluge of proposals for improving the program and successfully siting a repository followed. Common analyses of what was wrong with DOE's program focused on technical uncertainties and revising site selection criteria. Some suggested scrapping the entire site selection program and starting over. Others advocated implementing information sharing and techniques for mediation and negotiation.

Chapter 8 closes with a discussion of the implications of this analysis for the siting of other hazardous or controversial facilities.

Recommendations for revising the DOE program address only the symptoms, while conflicts caused by underlying stresses within U.S. politics and economics are ignored. Opposition to the repository siting program cannot be simply dismissed as a case of the NIMBY—not-in-my-backyard—syndrome. In the end, the failure to understand that locational conflict involved basic issues of political organization, participation, and power produced short-term attempts which addressed policy implementation problems but neglected the underlying long-term causes of such conflict. Some may suggest that because conflicts over the siting of hazardous facilities reflect basic tensions inherent in society, they are unresolvable—or that strong-armed federal preemption is inevitable. However, at this point in time, such a conclusion is premature. Our experience with these issues, and alternative attempts to resolve locational problems, is slight. We do not yet know whether this society can invent the conditions from which alternative approaches to these problems can emerge and be given a fair chance to succeed.

SITE UNSEEN

CHAPTER 1

▼

A PERSPECTIVE ON THE POLITICS
OF NUCLEAR WASTE

In his farewell address President Dwight Eisenhower
warned about the threat to democratic society from "the
military-industrial complex." This powerful coalition included not only weapons manufacturers and the armed services but also portions of academia, the scientific community, organized labor, and key supporters in the U.S.
Congress. As Eisenhower saw it, a strident ideology of national security and anti-Sovietism had been used to produce
"a still bigger defense industry and hence a still bigger political constituency in support of weapons development. This,
in turn, strengthened those elements of the Congress that
automatically endorsed any weapons-development program
and tipped the congressional balance of power still further
in that direction."[1] Eisenhower was no social theorist, but
his observations on the military-industrial complex suggest
some key insights into the workings of American society
and ultimately the politics of nuclear waste. In the first
place, the boundaries marking the state, economy, and civil
society are indistinct. Second, political power cannot simply
be equated with the institutions of government. Rather,
power is the demonstrated ability to shape ideology and scientific analysis and to dominate political agendas and decision making. If Eisenhower saw a Leviathan, it lived as
much in Cambridge as in Washington, D.C.

An examination of high-level nuclear waste disposal

might begin by considering the existence of a nuclear-industrial complex analogous to the military-industrial complex. To begin, one can reconstruct social forces and institutions which shaped politics, science, and people's thinking about the topic; and, consequently, shaped their views of the problem and its solutions. This chapter, then, is an attempt to define the network of relationships which existed among institutions of the state, economy, and civil society which together constituted a complex labeled here as "the nuclear establishment."[2] As used here, establishment simply means an orderly, relatively stable, and durable set of relationships —a recognized membership with common viewpoints and mutual privileges. It suggests "staying power" and a resistance to change. And while there may be internal stresses and conflicts (the DOE and the utilities don't always agree), "establishment" implies the ability to resolve them before they endanger the larger goals shared by members. By its nature an establishment is, therefore, exclusionary—identifying outsiders who are either critical or impassive to its operations.

The Nuclear Establishment Defined

More specifically, the nuclear establishment refers to a relatively stable set of relations among members of groups and institutions that promoted and benefited from the development of nuclear power and other nuclear technologies. A brief description of these groups and some of the key relationships among them can be found in table 1. In reading table 1, one would be mistaken to view these entities as a rigid set of completely consistent, immutable interests and institutions. The term *establishment* suggests a broad set of social, political, and economic relations that affected the development of nuclear waste policy and extended far beyond the federal legislative and executive branches of government. Within a state or local government, or even within the DOE, there may be varying support or opposition for the interests of nuclear power. For

TABLE 1

THE NUCLEAR ESTABLISHMENT

Entity	Roles and Relationships with Other Members
U. S. Department of Energy (DOE)	The key federal agency promoting the commercialization of nuclear power and nuclear research, including research on radioactive waste disposal; employing an extensive network of contractors, consultants, and national laboratories; a major contributor to state and local economies—e.g., especially those hosting the DOE nuclear reservations and DOE's operations offices; the agency responsible for the production of high-level waste from nuclear weapons plants and for the production of enriched uranium for reactor fuel rods.
Nuclear Regulatory Commission (NRC)	Initially composed of staff from the disbanded Atomic Energy Commission, the NRC is responsible for regulating nuclear power and nuclear waste disposal to assure public safety. It performs inspections and sets standards but has little money for research.
Selected members and committees of Congress	Originally, the Joint Committee on Atomic Energy promoted nuclear power, but later oversight was distributed among numerous committees. Individual state delegations vary in support or opposition. Congress oversees DOE funding but individual delegations and their constituents may receive the benefits of DOE-initiated projects.
DOE contractors for nuclear programs	DOE personnel actually perform relatively little of the agency's work as most work is done by contractors. For decades DOE contractors such as Bechtel, Du Pont, Rockwell, Battelle, and Westinghouse have provided management and engineering services to DOE and nuclear utilities. These contractors managed and operated federal facilities such as the Hanford Nuclear Reservation and the Rocky Flats nuclear weapons plant. DOE contractors may also include researchers in universities. Universities have received DOE research funding and, in turn, have furnished DOE, NRC, and nuclear industries with engineers and technicians. Universities may also run national labs, such as Argonne.

Nuclear industries	Industries which keep the nuclear fuel cycle running include uranium mining and milling companies (which have included energy companies such as Exxon); chemical companies (such as Kerr-McKee) that produce the uranium hexafluoride used in the fuel enrichment process; the fabricators of fuel rods and hardware such as spent fuel casks and reactor vessels; and companies providing nuclear engineering services. Their clients include electric utilities with nuclear power plants.
Nuclear utilities	The consumers of nuclear technology and services, utilities are also engaged in the day-to-day management of spent fuel and other high-level waste until another solution is found. Included in nuclear utilities are government agencies who are part owners of nuclear power plants. Historically, the utilities have assumed that disposal or recycling of spent fuel rods is a federal responsibility.
Selected local government entities	Some local groups and governments have been vociferous supporters of nuclear industries, utilities, and federal nuclearfacilities—soliciting such industries and lobbying state and federal legislatures for their continued operation, with local economic development being a major concern.
Selected state government agencies	Some states' governments and/or agencies have also promoted development of nuclear power and related industries—for example, the state agency that promoted spent fuel reprocessing plant at West Valley, New York. At times, states with uranium industries, such as Wyoming and New Mexico, may also be counted as participating in the nuclear establishment.
National labs	National labs, such as Sandia and Oak Ridge, are important DOE contractors. They have provided much of the research and computer modeling services that are used in DOE waste management programs.

example, while some state agencies championed nuclear power, others headed attempts to impose regulations on nuclear power plants which were stricter than those sanctioned by federal regulators. The key is to recognize the various efforts made by these members to create and promote nuclear power and manage high-level waste.

This broader set of relationships can be examined in terms of geographical and historical perspectives of space and time, and social theorists' perspectives on the horizontal and vertical organization of society (fig. 3). Taking each in turn, the following perspectives are used to direct this analysis:

1. The nuclear establishment is found at all geographic levels of politics, ranging from the local to the international. It incorporates members and groups within state and local governments who support, promote, and benefit from nuclear industries. The nuclear establishment may be expressed as a regional coalition of jurisdictions, a national coalition of industrial interests, an agency within the na-

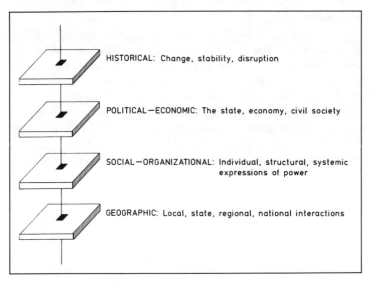

FIGURE 3. Contextual analysis.

tional bureaucracy, or as a sympathetic committee within Congress.

2. The nuclear establishment is historical and dynamic. Nuclear waste politics is more than a set of political actors engaged in a game of calculating and trading costs and gains. Over time, stresses among its members, as well as changes in its environment, challenged the establishment's ability to adapt and maintain control over its environment.

3. The nuclear establishment cuts across the horizontal organization of society—the institutions of state, economy, and civil society. The categories of private sector (economy) and the public sector (state) are often blurred and permeable. Private entities sometimes act as agents of the state, and public entities often assume the role of entrepreneur. Like Eisenhower's military-industrial complex, the nuclear establishment draws from many and diverse institutions— from the academic community of scientists and experts; national labs, private and public corporations; private and public electric utilities; mining, engineering, and consulting companies, government contractors, unions, and industry lobby groups.

4. The nuclear establishment cuts across the vertical organization of society—the individual, structural, and systemic. Individual political actors, such as legislators, take up the cause of nuclear power or the interests of members of the nuclear establishment. Actors and institutions set agendas which determine what issues will receive attention. At its broadest level the nuclear establishment is reinforced by social arrangements and structures which strengthen ideologies about social priorities and norms of political behavior.

Taken together, these four perspectives—the spatial, historical, vertical, and horizontal views of social organization —offer an alternative view of the politics of nuclear waste. Separate, each offers only partial insight. At worst, the adoption of one over the other leads to the predicament of blind men theorizing about the elephant. Narrow theoretical perspectives can, as one author observed, reduce the complexity of social, political, and economic relationships to little more than a cartoon version of reality.[3] However, taken together, these perspectives contribute to a more comprehensive explanation of the context out of which nuclear waste emerged as a topic of national debate. With its origins understood, one can understand how the nuclear establishment shaped the issues and political solutions to accomplish a reassertion of its own power and authority.

A Contextual Approach

The concept of a nuclear establishment, and the approach suggested for analyzing the politics of nuclear waste disposal, may be described as a "contextual orientation." The roots of this orientation are found in diverse sources of political and social theory. It grows out of dissatisfaction with the extremes of structuralism and behavioralism, Marxism and pluralism. In their examination of the three major theories of the modern state, Alford and Friedland recognize that "Each perspective has something to offer to the understanding of the state: The pluralist perspective contributes to a partial understanding of the democratic aspect of the state; the managerial perspective contributes to an understanding of the state's bureaucratic aspect; and the class perspective helps explain the state's capitalist aspect."[4] Their interest is in developing a synthetic framework capable of rescuing and integrating the major contributions of each perspective. Alford and Friedland propose that an adequate theory must incorporate three levels of analysis: the pluralist emphasis on the individual; the managerial on the organization; and the class perspective's emphasis on the systemic.[5]

The contextual orientation recognizes the social and historical context within which political action takes place. The purpose of the contextual analysis is not to discover the iron laws of history; rather, "The essential purpose is to enable the policy analyst, and hopefully the decision-maker, to find his way in the complexities of the total situation in which he operates."[6] As used here, contextual orien-

tation is part of an attempt to account for the complexities
of the total situation by linking the historical, spatial, social, and economic contexts of nuclear waste politics. It attempts to consider the different forms and dynamics politics
assumed at different locations and at different times. Hopefully, the contextual orientation demonstrates that analysis
is enriched by applying the multiple concepts and levels of
analysis. Before moving on to a discussion of the nuclear
establishment itself, additional discussion is needed on the
four perspectives underlying this approach.

A historical perspective

The argument developed over the next six chapters follows the historical development of nuclear waste policy.
Discussion begins with the glory days of nuclear physics,
the bandwagon years of nuclear power, and continues
through the challenges of the 1970s to the NWPA and the
conflicts of the 1980s.[7] Throughout this study, we are concerned with change; in particular, why was the nuclear establishment able to resist changes that over time would
have limited its power or disbanded key institutions (such
as the DOE) and their control over nuclear waste management? On a parallel track, one can follow the changes in
the U.S. economy and politics which, while they disordered
and threatened the nuclear establishment, were insufficient
to destroy it. What were the roots of the nuclear establishment's ability to successfully counter outsiders' attempts
to overhaul nuclear waste management?

If the historical context is recognized, there is little sense
in beginning to answer such questions with an analysis of
congressional debate over the 1982 Nuclear Waste Policy
Act. Descriptions of the behavior of politicians, voting records, the tactics, resources, and strategies used by lobbying
groups reduce explanation to an ahistorical set of policy
games. Neither is a legalistic interpretation of the NWPA
as the will of Congress the interest here. More important
is what that legislation said about political and economic

relations which constrained and channeled congressional debate and what it said about the future politics of nuclear waste. By examining the development of the assumptions informing its creation, nuclear waste policy is placed in the context of historical power relations and institutional priorities. Thus, the reader is asked to consider the following hypothesis: the NWPA. was not a break with the past; it set no new administrative gears in motion but merely confirmed existing powers, priorities, and practices.

Permeable boundaries: A political-economic perspective

Some theorists have argued for a view of the state "both as an independent entity with its own functions and goals and an integral component of the set of power relations which compose capitalist society."[8] In reality, the American state, far from being a monolithic entity, is organizationally and geographically fragmented. Besides legislative, judicial and executive institutions resident at the national, state, and local levels of government, the state encompasses a body of law, differing political practices and ethics, and ideologies of capitalism, democracy, and the role of the individual in society.

The actual boundaries among state, economic, and civil institutions are often unclear and permeable. Quasi-government institutions, corporations, and economic ventures such as the TVA and government mortgage corporations operate much like private corporations. Consider that attached to all levels of government are independent citizen advisory boards, councils, and commissions. There are intellectual institutions such as national research labs, local schools, and state universities to advise policy makers and shape political opinion about the causes of social problems. Individuals move between government agencies and working for industry.

In wartime the goals of the various state institutions may appear unified and harmonious. Most times, however, they are competitive and rarely coincide. The interplay among institutions of the state is a constant process of challenging and reorganizing the distribution of authority resident in federal, state, and local jurisdictions.[9] Conflict between national and lower levels of the state may be precipitated by attempts of national institutions to impose an externality, such as a repository, on a local jurisdiction. Such conflicts often focus on the interests of interstate commerce and industry versus the protection of public health, and accompany attempts by national institutions to preempt local authority (or vice versa). However, it would be too easy to consider such conflicts as simple cases of the local versus the national state, or of the national state's domination of the local state, or of the politics of production versus a local politics of consumption.[10] In the United States, state governments can successfully challenge the authority of the federal government. At other times and locations, the interests of some local and national institutions may coincide. Support for a federal policy is often regionally expressed with local jurisdictions in one part of the country favoring it and those in another part of the country opposing it. The history of nuclear power contains too many examples of regional, state, or local support for the nuclear establishment to draw nuclear waste disposal as a simple conflict between the central and local state.[11]

The political-economic perspective employed here considers that institutions of the state or economy cannot be treated as autonomous entities or uniform sets of interests. State revenue depends upon the health of the economy. Industry provides information to justify state intervention and decisions affecting the economy. Government regulation depends upon the cooperation of economic institutions (or may be solicited by those institutions) and members of civil society accepting the legitimacy of state authority. Through contracts and cooperative agreements, economic institutions may become de facto agents of the state. The state promotes and protects nuclear industries through research and development, regulation, and government insurance policies. It is even illegal for some industries, such as weapons manufacturers, to operate apart from the state. Capital can be both private and public. The nuclear economy could not function without huge state investments in capital projects such as federally owned uranium enrichment facilities.

A social-organizational perspective

Just as the nuclear establishment cuts across the horizontal boundaries of the state, economy, and civil society, it also cuts across the vertical dimension of social organization and expressions of power—the individual, organizational, and societal.

At the individual level, power is defined as control of political resources and the ability of actors to successfully negotiate favorable policy decisions, such as a bureaucrat's discretion over the geographic allocation of jobs and project funding, the power to favor some districts over others and, in turn, influence a legislator's support for the agency's budget or policies. Legislators who promote an agency may be rewarded with agency spending in their district or other actions intended to benefit local constituents. But local pork barrel politics does not explain all political actions and motivations. Representatives, lobbyists, and other political actors may use their power toward ends which have no obvious benefit for their constituencies; for example, their strength may be used to establish "credit" with other powerful figures, to maintain credibility with various constituencies, or to attract media attention.[12] Congressional supporters of nuclear power, for example, did not always represent states with commercial nuclear reactors; but the economic benefits from other nuclear facilities, such as federal research centers, or other political trade-offs were often at stake. Over the years key allies within Congress were instrumental in maintaining and protecting the nuclear establishment from political attack and dismemberment. Nevertheless, to look only at the situational level, and view power as a win-loss statement of a political group or politician, is to commit the pluralists' error of missing the fact that horse trading occurs within an institutional structure which sets the rules of the trading pit. Bachrach and Baratz, in their critique of the pluralist model, observed that "power is also exercised when 'A' [read a member of the 'nuclear establishment'] devotes his energies to creating or reinforcing social and political values and institutional practices that limit the scope of the political process to public consideration of only those issues which are comparatively innocuous to 'A.'"[13]

Thus, the power of the nuclear establishment is also structural—associated with the organization of institutions, procedures, and the command of intellectual resources. Organization defines a chain of command and responsibility which controls outsiders' access to decision makers, as is the case in which a citizen group may only speak with the public affairs officer. Loyalty to the institution is promoted and its members are socialized into an ideology of social priorities and institutional practice—what Secretary of Energy Watkins called the "DOE culture."

Power also resides in an organization's network of external contacts. Ties among members of the nuclear establishment created dependency. Frequent exchanges of goods, services, and personnel produced a stable network of mutually supportive institutions which reinforced each participant's goals and programs, and which were difficult to enter or disrupt. At the institutional level, power also originates in a fundamental property of bureaucracies—the creation of procedure. All institutions, whether state, economic, or social, define procedures that limit opportunities and avenues for outsiders' participation in the institution's planning and problem solving. Procedure outlines acceptable methods for recognizing problems. Very complicated procedures, requiring specialized navigational (usually legal) skills, discourage political activity and confine it within institutionally controlled channels. Those that fail to follow institutionally prescribed procedures may find themselves or their issue ignored. Situations which fall outside of standard procedures are usually judged to be beyond the scope of an institution's charter; therefore, procedural rigor may be used to limit institutions' responsiveness to outsiders' demands. At the same time, vague procedures can leave institutions with great discretion over what goals will be pursued and how they will pursue them. A lack of defined administrative procedures gives an institution an extraordinary amount of freedom to define key technological and locational issues.[14] Therefore, evidence of the power exercised by members of the nuclear establishment was found not only in individual policy decisions but also in nondecisions, that is, long-standing inattentiveness to issues raised by their opponents.

Another form of institutional power is access to, and the monopoly of, intellectual resources—technical expertise, data processing and information management, professional opinion, and research capability. Federal agencies responsible for nuclear power fostered and supported a vast network of technical expertise through their financial support of nuclear physics and engineering departments, national and private laboratories, private consulting and engineering firms. The livelihood of many organizations thus became tied to continued support for nuclear technologies. Technical expertise may define the language of political debate as organization limits the targets of political activity and procedure channels it. By adopting technical definitions of issues, the language of political discussion becomes specialized. Problem-solving criteria are removed from public debate, and the selection of solutions appears objective and mechanical. Defining an issue in strict, technical terms also defines an exclusive membership qualified to participate in political discussion.[15]

At the third level, societal or systemic power is control over widely held beliefs, concepts, modes of thinking, and expectations of political power or powerlessness. To understand this form of power, Gaventa says one must "focus upon the means by which social legitimations are developed around the dominant, and instilled as beliefs or roles in the dominated. It may involve, in short, locating the power processes behind the social construction of meanings and patterns."[16] At this level, power circumscribes conflict by enforcing paradigms and assumptions which structure thinking about politics, the possibility of social change, the optimal solution, and actions in the local or national interest. Technical expertise can be used, for example, to create and enforce a definition of the optimal location for a repository and then develop the mechanics for producing such a decision. The engineers' optimal solution may become equated with the best or socially desirable decision. Proponents may justify a controversial facility such as a repository by linking it to widely supported (but vaguely defined) values such as national security and economic development. In one case, local acceptance for a nerve gas facility was due in part to the locals' belief that they were doing something for the nation.[17] In such a case, opponents of the facility may then be labeled as a trouble-making minority, unpatriotic, or antigrowth.

The power to define political processes not only bars challengers from decision making; more important, it can create experiences of repeated failure for opponents of established institutions that evolves into an ideology of failure and indifference.[18] It's the "you can't fight city hall" idea, or, as so may Nevada residents seem to say, "Let's face it, the dump is coming here whether we want it or not. So we might as well get on with it." Pervasive quiescence is never likely to characterize the politics of nuclear power or nuclear waste. Nevertheless, one must still consider how the nuclear establishment affected thinking about energy and national security; the safety risks and benefits of nuclear technologies; and beliefs about political conflict and the motivations of its political opponents. As we shall see, the nuclear establishment promoted the view that political processes were inefficient and only produced unproductive conflicts which hindered progress toward a solution. The nuclear establishment felt that the conflicts would be settled once the public was properly educated to accept that nuclear waste and its repositories could be managed safely.[19]

A geographic perspective

In practice politics is spatial. Constituencies, jurisdictions, issues, and arenas for political debate all have a geographic dimension. The state is fragmented into national, regional, and local jurisdictions. In the economic sector, nuclear utilities serve regional markets and answer to state public utility commissions (PUCs). Nuclear energy is tied to international uranium mining, engineering, and chemical corporations. Civil society includes national labor organizations; local activist groups, clubs, and chambers of commerce; national environmental groups—with and without local chapters; the local and national media. Thus we must recognize that the nuclear establishment and its opponents operate in a variety of contexts and not simply at the level of national policy. In the case of nuclear waste disposal, the following three spatial expressions of politics are particularly evident: the issue arena; local challenges to the national state; and spatial disparities in the distribution of costs and benefits.

In the politics of nuclear waste there is an on-going tension as to the geographic *arena* in which issues will be addressed. The national state, with the support of national economic interests, may attempt to preempt or limit the regulatory authority of state and local jurisdictions, for example, by invoking the interstate commerce clause of the U.S. Constitution to preempt state/local regulation of nuclear waste transportation. National groups or institutions may attempt to shift a problem onto the local level—by making emergency response to a transportation accident involving hazardous materials a local responsibility, for example. Similarly, the implementation of national environmental policies is often played out at the state/local level. A state may ignore national air pollution standards to encourage industrial development. On the other hand, states may question the federal government's commitment to radiological safety and develop standards stricter than the federal ones. In times of crisis, states may attempt to shift the burden of relief onto the national state, as in the case of flood disaster relief. Tensions also exist between local and state governments. Local economic problems often are blamed on neglect by state policy makers. An example of shifting an issue or crisis to another spatial level is found in the case of nuclear utilities that blame the lack of spent fuel storage on congressional inaction or obstructionist states.[20]

Once defined in location-specific terms, the political significance of an issue and the range of participants are reduced. Local challenges to national policies and programs may be vociferous but have inherent weaknesses which reduce their chance of victory. Their interests and base of support are parochial, making a link with national groups difficult. Regional or national opponents of community activists may point to their provincial nature to discredit them as self-serving and obstructing the national interest. They have limited resources and staying power since they are often organized around a single issue.[21] Finally, a political victory may not resolve the problem but simply export it to another area. However, from another point of view, deferring issues to the local level means that conflict can be circumscribed, planning procedures simplified, and transaction costs reduced (which is advantageous when industry is required to finance state actions). The origin of a problem in national policies and structures may be largely forgotten once resources are dedicated to producing a local, issue-specific solution. Thus one must consider how the power to adjust the spatial expression of the nuclear waste issues, and the forums in which they were addressed, altered political opposition and the effectiveness of challenges to the nuclear establishment.

State and local institutions are not, however, powerless to challenge federal agencies and large industries. Nuclear utilities, for example, are subject to a state's public utility commission and depend upon the cooperation of state governments and local police and fire departments to develop an emergency response plan necessary to obtain an operating license for a nuclear power plant. Uncooperative state and local governments can stop already built reactors from coming on-line, as occurred in New York and Massachusetts, for example. State and local challenges can disrupt national markets, prompting industry demands for federal policies that override such actions. National regulation provides industry with an economically advantageous uniformity which frees it from the anarchy of multiple sets of state regulations.[22] Thus conflict is likely to characterize any state or local attempt to regulate nuclear industries in accordance with local interests. Historically, members of the nuclear establishment have aligned with agencies who support federal preemption and oppose local initiatives; its opponents have generally supported state and local controls.

A third geographic aspect of the politics of nuclear waste is the spatial distribution of costs and benefits. The spatial concentration of benefits (or disbenefits), combined with the spatial segregation of beneficiaries from cost bearers, provokes political activism. Local opposition to a project builds on discrepancies—such as electricity from nuclear power plants being used in the East with the nuclear waste dump situated in the West. To counter claims of inequality, the nuclear establishment attempted to nationalize the benefits by asserting that everyone benefited from nuclear energy because it guaranteed the energy supplies necessary for a strong and safe America. On the other hand, if a repository was geographically isolated in a place such as Nevada, the negative side effects of it would be largely invisible to the rest of the country.

Geographical isolation may translate into political acceptability. A remote, rural population is unlikely to have the political power, resources, or expertise, to influence the distribution of externalities. Isolated from the national economy, a rural community may even view a repository as a desirable economic development. Even when addressed, disparities can be more cheaply offset at remote locations. Thus, financial compensation becomes a substitute for addressing existing political-economic structures as the cause of inequities. The resources to provide economic compensation or benefits therefore become another form of power.[23] As one example of this, a portion of the contributions to the Nuclear Waste Fund are used to compensate for local impacts caused by the repository program. But discretion to decide the legitimacy of claims and the control over distribution of monies rests within the U.S. Department of Energy's Office of Civilian Radioactive Waste Management. The autonomy and power of local institutions is lessened as they become financially dependent on national ones, further enhancing the power of the national state and those capable of influencing its policies.

Questions and Context

Building upon the concept of a nuclear establishment, this chapter discusses factors which affected the development of nuclear waste policy. Figure 4 portrays some of the links between the analytical perspectives highlighted in this chapter and the argument developed in the remaining chapters. As shown in Figure 4, the nuclear establishment consisted of more than nuclear utilities and the U.S. Department of Energy. The various cells show that its membership included allies within Congress, industries participating in the nuclear fuel cycle, nuclear scientists and engineers, contractors, industry organizations, and promoters of nuclear power within state, local, and federal institutions. The boundaries among the cells are permeable, emphasizing their interdependence and the point at which promoters of nuclear power and technologies cut across the boundaries of state, economy, and civil society. Rudolph and Ridley captured this permeability when they described electrical generation as "the most politicized business in America."[24] Nuclear utilities are supported by a complex mix of private and public, regional, local, and national industries.

Key political personalities, large institutions, and a social system based on economic expansion and the protection of private investment—corresponding to the expression of the individual, structural-organizational, and systemic levels of power—combined to foster and protect nuclear energy.

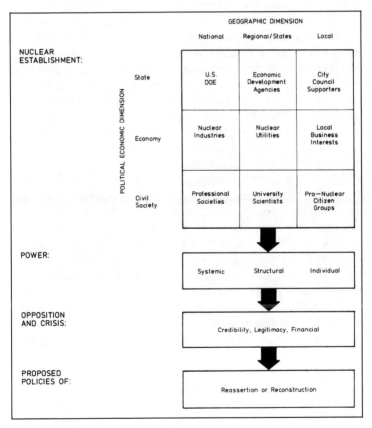

FIGURE 4. Contextual analysis of nuclear waste policy.

Support for the nuclear establishment extended from the politics of national security and national economic strength, down to communities' interests in jobs and economic development—one manifestation of the geographic dimensions of the conflict. Control over scientific and technical resources gave the nuclear establishment the power to craft decisions about the desirability and feasibility of nuclear technologies and policies. Thus, the conflicts over the siting a nuclear waste repository were much more than a problem of poor program administration, more than a site-specific dispute over externalities and compensation, more than a case of the not-in-my-backyard (NIMBY) syndrome. Throughout the history of nuclear waste programs one finds a tension between attempts to preserve the stability, authority, and continuity of the nuclear establishment and demands that it respond to a broader range of social and environmental concerns. Historical, geographic, political, economic, and organizational origins of the politics of nuclear waste must be understood if one is to appreciate the origins and nature of the locational conflicts.

The power to shape nuclear waste policy took a variety of forms. While powerful politicians negotiated compromises which resulted in the 1982 Nuclear Waste Policy Act, organizations and science shaped the topics and language of political debate. From this point of view, nuclear waste policy was not the engine that drove politics, but the product of political, economic, and social engines which drove the politics of nuclear waste.

Hopefully this analysis of the institutional and social constraints on political actions and debate does not lurch to the extreme caricature of political actors as mere agents of elites or special interests. Most important, if readers recognize the problem of nuclear waste disposal as a historical and political-economic construct, they will then recognize that no analyst can produce a "correct" solution to the question of what should be done with the by-products of a nuclear economy. The lessons learned from our experience with nuclear waste disposal offer a warning to others seeking quick resolution of complex problems and conflicts involving the disposal of toxic materials or the siting of other potentially hazardous facilities.

CHAPTER 2

THE NUCLEAR
ESTABLISHMENT

Most histories of nuclear power tend to be narrative descriptions of national institutions and policy-making. Many of them are expressly pro- or antinuclear.[1] Histories of nuclear waste are relatively uncommon. Most of these works offer descriptions either of waste disposal technologies or the administration of federal programs.[2] They also have not provided insights into the broader forces shaping the politics of nuclear waste. To begin such an analysis, this chapter takes a closer look at the origin and nature of the nuclear establishment. Relationships among the various groups described in table 1 could be analyzed in terms of the contextual approach described in chapter 1. However, because it would be tedious to document all such interactions, the nuclear establishment is analyzed here in terms of six major themes.

State Sponsorship

The nuclear establishment was built on the foundations of the wartime Manhattan Project. Government contracts for the research and development of the nation's first nuclear weapons favored large chemical and engineering firms such as Du Pont, General Electric (GE), Union Carbide, Stone and Webster, and a few large universities such as Columbia, Princeton, Chicago, Minnesota, and Berkeley. In the years following World War II, nearly all nuclear research was classified, which left little opportunity to develop commercial applications. As the potential commercial applications of nuclear energy became more and more evident, debate focused on two choices: (1) nuclear power should remain a federal monopoly and the equivalent of a nuclear Tennessee Valley Authority should be created to serve the dual purpose of producing electricity and raw materials for nuclear weapons; or (2) the doors should be opened to privatization. David Lilienthal, head of the TVA between 1933 and 1936, favored maintaining a state monopoly. However, his successor as chair of the Atomic Energy Commission (AEC), Lewis Strauss, championed commercialization. In 1953 Strauss was largely responsible for bringing together Westinghouse and the Duquesne Light Company to build the first civilian reactor to generate power on a commercial scale—albeit uneconomical and 90% paid for by the federal government. Between 1954 and 1966 the federal government funded numerous other demonstration reactors as part of its effort to promote commercialization. According to a 1961 AEC report, between 1946 and 1959 the commission invested $2.5 billion for reactor research and development, and nearly a billion more for construction of facilities, compared to industry's contribution of $650 million.[3]

The prototype commercial reactor at Shippingport, Pennsylvania, was actually a scaled-up version of a Westinghouse reactor developed for Admiral Rickover's nuclear navy. In general the reactors proposed for the commercial generation of electricity were scaled-up versions of military reactors. Rather than gradually scaling up reactor capacity and testing designs at each step along the way, the AEC favored rapid development. For GE and Westinghouse, short-circuiting the research and development process and recycling military reactor designs had several advantages: it permitted them to maximize their financial return from past work on military reactors for which the federal government had already paid, and they avoided sinking their own funds into the development of new reactor designs for the com-

mercial market. It also favored their continued dominance
of the nuclear reactor and engineering markets.

In the history of nuclear power 1954 was a key year. That
year the first nuclear-powered submarine, *Nautilus*, was
launched. The Atoms for Peace program had been announced only a few months before in President Eisenhower's address to the United Nations. Provisions in the
1946 Atomic Energy Act, which had kept reactor design data
classified and guaranteed a federal monopoly, were discarded. Most important, the federal government—but not
necessarily the Tennessee Valley Authority—was prohibited from entering the business of owning commercial reactors and using them to generate electricity. The federal monopoly of nuclear-generated electricity, which was feared by
the utilities, coal companies, and reactor suppliers, had been
averted. Utilities were given the lead in developing commercial nuclear power. Nevertheless, the turn toward commercialization was not a complete turn toward privatization.
The 1954 Atomic Energy Act set up the Reactor Demonstration Program which would subsidize commercial construction so generously that some nuclear promoters condemned
it as an effort to "force feed atomic development" at the
taxpayers' expense.[4] It was defended as enhancing U.S. industries' ability to compete against the Europeans, especially the British Atomic Energy Authority, for new markets. The rivalry was rooted in the early development of
nuclear physics that later extended into the uranium enrichment and fuel fabrication industries as well. The federal
government offered to provide fuel for these first commercial reactors and, through its price-fixing policies, produced
a boom in the uranium mining industry. Subsequent amendments to the 1954 act permitted the AEC to finance reactor
research and development costs and waive other costs. The
act directed the AEC to assume responsibility for the management (for example, reprocessing) or disposal of spent fuel
from commercial reactors—a seemingly insignificant responsibility at the time.

Even as the development of nuclear power was being

turned over to private industry, the state continued to play a key role in fostering that development. The 1954 act maintained federal preemption of the authority to regulate the nuclear industry, freeing utilities from pesky state regulations that could interfere with their operations. The AEC's relationship with Congress was generally benign. The Joint Committee on Atomic Energy (JCAE) monopolized congressional oversight and, with few exceptions, supported and promoted the AEC's mission. The AEC was both regulator and promoter of atomic energy and for years no other federal, let alone state, entity threatened its authority. Growth of the industry and advancement of nuclear power was accomplished by administrative fiat. The goals and policies of the promotional campaign were never spelled out in legislation. Similarly, the nuclear waste program would evolve for thirty years without explicit legislative direction until passage of the 1982 Nuclear Waste Policy Act. One observer summarized the close-ended nature of the politics of nuclear power as follows:

> Starting with the premise that atomic technology should be developed, introduced, and used, the JCAE and the AEC have usually cooperated symbiotically to resolve policy issues behind the scenes before legislative proposals can reach public view. By the time these proposals emerge, it can be said that they are "noncontroversial," and they usually breeze automatically through Congress [as part of the budget and appropriations process]. Benefits of nuclear technology have usually been assumed, while risks have been ignored or dismissed in boldly optimistic terms.[5]

When demands for public access and more oversight of nuclear programs were finally heard, the organizational structure of the nuclear establishment became more complicated. In 1973 the AEC was disbanded. Its promotional and regulatory functions were divided between the newly created Nuclear Regulatory Commission (NRC) and the Energy Research and Development Administration (ERDA). The breakup of the Joint Committee on Atomic Energy fol-

lowed in 1976. Where once one joint congressional committee had oversight of nuclear programs, now up to a dozen
committees had jurisdiction. The creation of the NRC produced a new administrative apparatus to regulate the commercial nuclear industry (but not military or experimental
facilities) for the protection of public health and safety. The
promotional function was left to ERDA and its successor,
the Department of Energy (DOE). But institutional continuity was still assured since the staffs of the NRC and the
energy agencies came largely from the AEC. Free traffic in
employees between the agencies and the nuclear industry
continued. Even following these changes, and President Carter's appointment of two NRC commissioners critical of the
nuclear industry, federal budget allocations left no doubt
that the promotion of commercial nuclear power would
continue.

The Nuclear Economy

The links between the state and economy are especially
complex in the case of nuclear power. The electric utility
industry is itself a mix of investor-owned companies, federal, local, regional or state systems, and electric coopera-
tives and interstate power pools.[6] Despite the atomic energy
acts, the state is directly involved in the production and
distribution of nuclear-generated electricity. Utilities such
as the TVA rival investor-owned utilities in size and revenue. Cities, small towns, and rural counties own shares in
multi-billion-dollar nuclear power plants. Interstate power
pools bring together investor and state-owned utilities into
regional service agreements which further reinforce the ties
between the state and industry. Thus, portions of every level
of the state, from the national to the local, were united in
the nuclear fuel cycle (fig. 5) and have some interest in how
the spent fuel disposal problem is solved. Probably in no
other industry have the boundaries between state and economy been so blurred.

When the AEC first attempted to promote the commer-

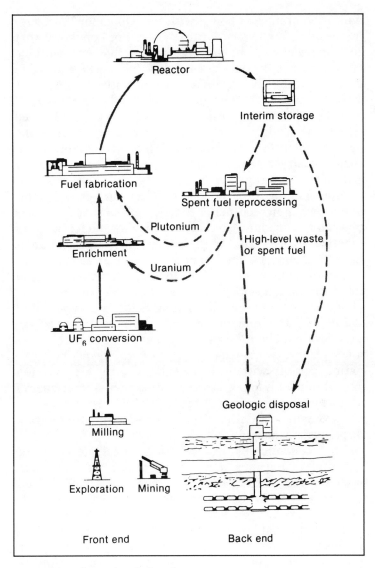

FIGURE 5. The nuclear fuel cycle.

cialization of nuclear power with the Atoms for Peace program it had few takers. Utilities were not convinced that nuclear power could be economically feasible. At least three key uncertainties—accident liability, fuel costs, and waste disposal—had to be eliminated before utilities could trust that nuclear power was economical. Federal intervention removed each barrier in turn. The Price Anderson Act limited industries' liability in the event of a reactor accident; the AEC offered utilities the benefit of a federal uranium enrichment industry, and agreed to accept utilities' spent reactor fuel for reprocessing.

From 1945 to 1970 demand for electricity grew at rates which have not been matched since; electric power usage grew at a compound annual rate of 7.8% and peak load growth averaged 8.1%.[7] With the major barriers to commercialization removed, nuclear power was an attractive means for meeting this extraordinary growth in demand. Not only did it promise freedom from uncertain fuel supplies and labor costs due to unrest in the coal fields; it also promised to free utilities from unpredictable railroad coal haulage rates and rapidly increasing oil and gas prices. Nuclear power plants also offered utilities a way to expand their generating capacity without being subject to the air pollution control standards imposed on coal-fired plants—particularly advantageous in portions of the upper Midwest, the Northeast and Southern California, which had difficulty meeting air pollution standards.

The economic significance and political power of the large utility companies was national; they controlled a form of energy critical to nearly every economic activity. Their research and lobbying groups, such as the Edison Electric Institute, the Electric Power Research Institute, and the Atomic Industrial Forum (founded in 1953), participated in the formation of nuclear policy. Their profitability, however, was still dependent on rate-setting by state public utility commissions, and the commissions' determination of acceptable costs to be passed on to the consumer. Eventually, state regulation would become an important constraint on

the technological, locational, and financial options for dealing with a growing inventory of spent reactor fuel. But in
the early years of nuclear power, with the promise of federal
relief from nuclear waste disposal costs, utilities had no reason to view waste disposal as an important issue.

Local Support

Until the 1970s, state opposition to nuclear technologies,
let alone nuclear waste programs, was timorous and largely
ineffective. Acquiescence was not complete, however, and
some projects were stopped. For example, in 1962 Consoli-
dated Edison was prevented from building a seven hundred
megawatt reactor in the middle of Queens. In the 1950s the
construction of the Fermi-1 breeder reactor between Detroit
and Toledo was opposed by the United Auto Workers. The
Fermi reactor is interesting because of the AEC's blatant
attempt to force its development, even if it meant disregard-
ing concerns about its risks raised by advisors to Congress
and the AEC's own advisory committee on reactor safety.
Though the breeder reactor was eventually built, it produced little power. But it did produce one of the country's
most serious reactor accidents. On the positive side, as a
result of its opposition, the United Auto Workers were successful in procuring a requirement that future AEC hearings
be open to the public.[8] However, through the 1960s and
1970s, even as radiological accidents, leaks, and health problems were publicized, many state governments continued
to voice their support for nuclear technologies. The state
of Washington elected as governor Dixie Lee Ray, former
head of the AEC. In at least one case a state government
assumed a role similar to the AEC and promoted the development of nuclear industries. The New York State Energy
Research and Development Authority promoted the development of a spent fuel reprocessing and high-level waste
facility at West Valley, New York, with the expectation that
it would attract other businesses to the area. Then Governor
Rockefeller saw nuclear energy as the desirable high-tech

industry of his day. Other state agencies became involved in the promotion of nuclear power: the Power Authority of New York, the largest nonfederal public power organization in the United States, invested heavily in nuclear power plants. Testifying on behalf of the Public Power Association (an organization of 1,750 municipal, state, and local power systems), Thomas Frey promoted the national benefits of nuclear power plants. Responding to proposals that nuclear waste disposal should be funded by the users of nuclear-generated electricity, Frey said:

> This does not mean, however, that such facilities only benefit the ultimate users, the consumers of electricity. For example, the low cost nuclear energy produced by the Power Authority has served to maintain and encourage commerce and industry in the State of New York. Therefore, a nuclear waste program not only benefits the facilities' users, it is a vital element in the continued economic stability of this country and its efforts to reduce use of foreign oil.[9]

Local governments also actively promoted nuclear power. Towns and cities all over the United States solicited nuclear industries, ranging from uranium mining to reactor fuel fabrication. Eventually, some sort of nuclear facility which produced nuclear waste could be found in at least thirty-nine states and near most major metropolitan areas.[10]

But state and local acquiescence was never comprehensive and changed as the mass media publicized cases of mismanagement and environmental hazards at waste facilities and nuclear reactors. In the landmark *Calvert Cliffs* decision, the court overruled the AEC and required an environmental impact statement under the 1969 National Environmental Policy Act (NEPA) before the AEC could permit a commercial reactor. NEPA thus gave the states and citizen groups a powerful tool with which to force entry into the decision-making process. Other environmental legislation and programs for state implementation of legislation, such as the Clean Air Act of 1970, led to the development and expansion of state departments of health and environmental quality. Some state agencies developed a base of experience, expertise, and sophisticated organization that eventually permitted states such as California and Minnesota to mount effective challenges to federal nuclear regulations. In the late 1960s, federal attempts to site nuclear waste repositories and conduct exploration work for such facilities brought home the realization that sooner or later states would have to confront the issues of spent reactor fuel and high-level waste disposal.

The Community of Experts

Another component of the nuclear establishment that would significantly affect nuclear waste policy was the growth of the an extensive network of experts and institutions involved in nuclear research and development. From the beginning, nuclear research was assumed as an activity of the federal government; initially, the research's military and strategic value was used to justify the secrecy. Later, when much nuclear research was declassified by the 1954 Atomic Energy Act and the Atoms for Peace program, an already extensive network of national labs and government contractors received the additional directive of aiding the commercialization of nuclear energy. The precedent had been set. Historically, utilities contributed relatively little to the research and development which supported commercial nuclear power. Thus, the national state had a near-monopoly on key forms of technical expertise, especially those related to areas of exclusive federal responsibility such as underground nuclear waste disposal.

Through their sponsorship and professionalization of this new industry, educational institutions developed close ties with federal agencies, government contractors, and the network of national laboratories engaged in nuclear research. Departments dedicated to the development of nuclear technologies were established; university programs in nuclear engineering fed the demand for nuclear scientists and engineers; Ph.D.s held important administrative positions throughout the nuclear establishment. In turn, federal research money flowed into universities, and the Atomic Industrial Forum was founded as an alliance of universities and corporations. The backing of institutions of higher learning lent credibility to the AEC's mission of promoting nuclear power and to early assertions that waste was a non-issue and could easily be disposed of by burying it in salt formations.[11] Scientists whose research raised too many questions about the safety of nuclear power and federal nuclear waste management practices were ostracized by the national and university research communities.[12]

The distinction between public and private sectors was not readily apparent in the management of federally owned nuclear research, production, and experimental facilities. Contractors such as Dow, Union Carbide, Westinghouse, Allied Chemical, Rockwell, or Battelle were hired to manage federal programs and government-owned facilities. Contracts lasted for years and close relationships developed between contractors and federal agencies. The expertise of these companies was tapped to fill key policy-making positions throughout the government. For example: past executives at Bechtel—the architect and engineer of over fifty U.S. nuclear power plants and reactors in South Korea and Japan—included President Reagan's Secretary of State George Shultz and Secretary of Defense Caspar Weinberger. Companies such as Rockwell had decades of experience and investment in federal facilities like Hanford by the time the Nuclear Waste Policy Act was being debated. (Later, however, charges of mismanagement were leveled at Rockwell and it lost the Hanford contract to Westinghouse). Protecting past investments in technologies, research, and expertise was a key consideration in defining an acceptable solution to nuclear waste problems.

Institutional Organization and Procedure

The organization and procedures of institutions within the nuclear establishment were characterized by technocratic forms of decision making, inconsistent tendencies toward fragmenting and centralizing of authority, an unwillingness to consider nontechnical problems, and confidence in the technical fix. In its early years, the AEC was independent and mostly insulated from public oversight. Nuclear energy was a subject commonly assumed to be too complex for politicians and citizens to understand. The organization and independence of the AEC was an affirmation that the experts were in the best position to make decisions and policies. One analyst saw confidence in technology and a disregard for nontechnological aspects of the problem as major themes found throughout the history of nuclear waste policy, which also characterized contemporary nuclear waste management. Writing in a 1978 NRC report, Metlay observed that these modes of thinking ultimately led to a situation in which

> the persistent faith in a technological fix has produced a myopic vision of the waste management problem. In theory, as well as in reality, the [institutional] boundaries of the waste management "system" have been severely circumscribed. This constrained view of what must be considered in designing a waste management system has resulted in a number of significant distortions.
>
> First. . . . Those who believe in a technological fix strive to eliminate the human factor—an element which, it is generally held, can only produce noise. Yet, time and time again, persons interviewed in preparing this report stated that the weakest link in a waste management system will be the human one.
> . . . .
> A second distortion . . . is the very high discounting of factors which may be affected indirectly by the system.[13]

Metlay goes on to describe how technocratic decision making served the interests of economic growth and efficiency, to the neglect of social and environmental impacts and concerns.

Over time, administrative attempts to promote the quick technological fix characterized repository programs. In the early 1970s, the AEC proposed Project Salt Vault at Lyons,

Kansas. The AEC refused to consider additional studies of the site because such studies would delay the project by two years. (Similarly, DOE refused to study granite in the first round of the NWPA repository program, because it could delay opening of the first repository by four years.) Eventually, this jump to the quick fix and the risks of premature decisions led to demands by environmental and other groups for closer oversight of the site selection process.

In the early years of the nuclear establishment, however, nuclear waste was viewed as a nonproblem. Confidence in the availability of a technical fix provided a rationale for separating the problem from the rest of the nuclear fuel cycle, making waste disposal irrelevant to decisions about reactor development, spent fuel reprocessing, or other segments of the fuel cycle. Other factors reinforced this neglect. A nuclear waste dump lacked the pork barrel qualities or the status of a research center or reactor testing station like the Idaho National Engineering Lab. Some members of the nuclear establishment, such as AEC chairman and later Secretary of Energy James Schlesinger, advocated exotic, high-profile technologies of transmutation and space disposal. In general, however, few were interested in the problem, and none championed the issue with the same fervor dedicated to reactor development. According to Metlay, Dixie Lee Ray, the archchampion of nuclear power, would "turn up her nose" when the subject was mentioned at meetings. Within member institutions of the nuclear establishment, no career ladder was associated with nuclear waste management and little money was allocated to it. Between 1961 and 1969 less than thirty million dollars a year of federal money was budgeted for it. No organization on par with the reactor development program was assigned responsibility for addressing waste disposal until after passage of the 1982 NWPA. Nuclear waste was truly the residual of the nuclear establishment.

Federal responsibility for waste management was fragmented among field and operations offices at numerous locations, including Hanford, Savannah River (South Carolina), and the Nevada Test Site. Until 1973 the internal allocation of AEC's budget was not even monitored by Congress. Each field office had wide discretion to set its own priorities and procedures, and different field offices promoted various versions of high-level waste disposal. Because they employed thousands of people and their contributions to state and local economies totaled millions of dollars per year, powerful coalitions in Congress sought to maintain this network of facilities and contractors. Energy officials in Washington, D.C., were relatively ineffective in controlling the management or operations of these field facilities—a problem that would later plague DOE's attempts to coordinate its high-level waste disposal program. A complicated array of contractors, subcontractors, and suppliers supported these facilities. Outside groups interested in monitoring them often found themselves dealing with contractors rather than federal officials.

Before NEPA and the NRC, decision-making procedures related to nuclear power and waste management were largely an internal matter. The methods or criteria used to reach a decision were often not apparent to observers outside the nuclear establishment. The lack of clear decision-making procedures permitted great latitude in deciding which problems and technologies to emphasize and which, such as nuclear waste, to neglect. Federal energy agencies were granted the discretion to adjust program priorities, timetables, decision-making criteria, and procedures virtually at will. Eventually, demands for set procedures, program goals, and decision-making criteria would become key issues in the nuclear policy debate.

Groups in civil society, such as the United Auto Workers, the Union of Concerned Scientists, and the Natural Resources Defense Council demanded, and in some cases eventually obtained, greater public access to the organizational network, procedures, and technical resources of the nuclear establishment. In time, decision making became a labyrinth of public hearings, environmental impact statements, decision documents, and procedural standards. The boundaries of each institution's authority and responsibility often remained unclear, producing conflicts over attempts to define or preempt them. For example, the organization of nuclear waste transportation became particularly complicated: the NRC approved routes, certified designs for shipping containers, and supervised loading at commercial reactor sites (DOE had authority over weapons and experimental facilities). The U.S. Department of Transportation (DOT) regulated hazardous materials transport; state and local governments were responsible for emergency response; and private carriers supervised driver training, vehicle maintenance, and actual vehicle operations.

Spatial Aspects of the Nuclear Establishment

The spatial organization of the nuclear establishment—crosscutting local, state, and national levels of institutional organization—produced a complex geography of politics which can not be drawn simply as a conflict between the national and local state. The nuclear establishment was able to affect the geographic distribution of facility-related costs and benefits and, thus, the scale of issues, participation, and conflict.

The issue of the geographical distribution of costs and benefits associated with nuclear power permeated the debate over nuclear waste policy. Obviously, most nuclear reactors and consumers of nuclear power in the United States are located east of the Mississippi River, although a significant number exist on the West Coast (fig. 6). Nuclear power was quickly adopted by utilities in the northeastern United States where generating electricity with oil-fired boilers was prohibitively expensive. Utilities serving metropolitan areas in the upper Midwest and the eastern seaboard looked upon nuclear power as nonpolluting (compared to coal-burning power plants). After the idea of siting large reactors in remote areas was abandoned in the 1960s because it was uneconomical and would send the wrong message to the public about the safety of nuclear reactors, they were built close

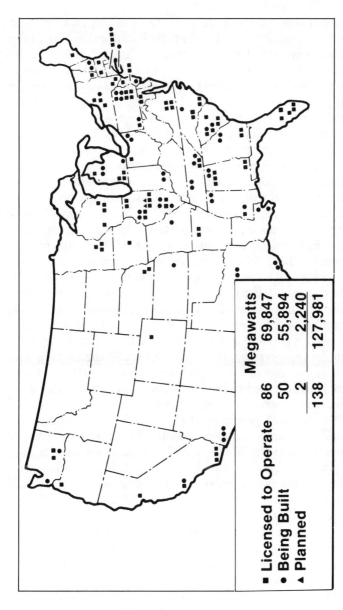

		Megawatts
■ Licensed to Operate	86	69,847
● Being Built	50	55,894
▲ Planned	2	2,240
	138	127,981

FIGURE 6. Nuclear generating capacity in the United States shortly after passage of the NWPA.

to the major metropolitan consumers of the East.[14] Consequently, inventories of spent fuel and other high-level
wastes were also located near urban areas. States targeted
for high-level nuclear waste dumps received an insignificant
portion of their electricity from nuclear plants—a geographical pattern even more clearly defined after potential salt
sites in the Midwest and potential repository sites in the
eastern United States were abandoned. Much of the high-
level waste from military reactors and reprocessing plants
was found in South Carolina and Idaho, which were also
considered unlikely sites for the first repository. Lack of
spent fuel and other high-level waste storage, contrary to
alarmists, was not a national problem but was concentrated
in specific states at specific locations. In addition, the problem was unevenly distributed among utilities with nuclear
reactors. The NRC essentially had no minimum requirements for reactor site storage other than what would be required to empty the reactor core in the case of an emergency.
Some utilities assumed that the federal government would
begin reprocessing spent fuel, and installed only small fuel
rod storage pools. When President Carter declared a moratorium on reprocessing, these utilities found themselves with
insufficient storage space. While nuclear waste policy acts
were being debated in the 1970s and early 1980s, these utilities claimed serious near-term storage problems. In the late
1970s the DOE predicted that by 1983 dozens of reactors
would run out of storage capacity. However, estimates of
when reactors would run out of storage space were revised
so many times that such projections lost credibility with
many members of Congress.[15] In reality, no utilities would
run out of storage capacity between 1980 and 1990, although
some were forced to invest in expanded storage at reactor
sites. But even if nuclear waste disposal was by no means
a crisis threatening the entire country, the possibility of reactor shutdowns or of turning reactor sites into de facto repositories was a powerful image which nuclear utilities
could capitalize upon in their attempts to force federal relief.

Congressional representatives from regions served by

utilities which faced shortage problems (such as the Southeast) took an interest in waste disposal legislation. Other representatives saw legislation as a way of removing waste from their home districts. For example, Representative Butler Derrick of South Carolina lobbied hard for passage of federal nuclear waste legislation. South Carolina was the home of a low-level waste dump, much of the nation's military high-level waste, and was once considered a possible host for a high-level waste facility. The inoperative Barnwell reprocessing plant was also in South Carolina. It held spent fuel in its storage pool and was sometimes eyed as an interim storage site until a repository was built. Jurisdictions with nuclear waste were anxious to have it shipped elsewhere. States with facilities suitable for temporary storage of spent fuel—including Illinois, South Carolina, and New York—promoted a remote repository site as the only feasible solution. Repository host states, pointing out the uncertainties of permanent disposal and the hazards of transportation, countered with proposals for temporary storage sites close to reactors. Utilities wanted both interim and permanent federal sites—seeing the former as a relief from the costs of expanding reactor site storage and a buffer from the uncertainties of repository development.

Originally, an eastern repository was viewed as necessary to achieve an equitable distribution of the negative impacts associated with high-level waste disposal. The regional approach, championed by Representative Morris Udall of Arizona, answered the perceived inequity of the eastern United States receiving nuclear-generated electricity while the West was forced to take the waste. By delaying consideration of sites for an eastern repository, DOE chose to concentrate impacts and thus state/local opposition in the less demanding political environment of a western location. Historically, agencies ranging from DOE to the General Accounting Office, along with many politicians from eastern states, favored repository locations with a history of nuclear activities which would "preferably [be] on federal land, including land where wastes have been stored or deposited."[16] Such locations held the possibility of reduced local opposition and deference to national institutions' dominance and control over decision making.

Conflict over the siting of a noxious facility is most often explained as the "not-in-my-backyard syndrome." The local concentration of negative impacts generated by a noxious facility, and the geographically diffuse nature of its benefits, are said to motivate local political opposition to the facility but indifference elsewhere. Numerous case studies of local opposition to noxious facilities, such as landfills and toxic chemical dumps, support this view.[17] However, after studying a case in South Dakota where local residents supported a low-level waste facility but the rest of the state opposed it, Shelley and Murauskas concluded, "the noxious facility locational conflict model appears to have little relevance to this conflict."[18] Such findings suggest the importance of recognizing that local and regional patterns of support or opposition to the nuclear establishment will be complex and inconsistent, demanding attention to historical and situational characteristics unique to that location.

State-level challenges to the nuclear establishment were complicated by divisive internal conflicts between large metropolitan areas and small towns which viewed a waste disposal facility as an economic gain. For example, a town in Washington near the Hanford facility solicited a repository as a desirable addition to the local economy (a local high school used the mushroom cloud as its logo), while other state residents vehemently opposed it. Examples of local support for high- and low-level nuclear waste facilities—in contrast to general state opposition—can be found in southeastern Utah; Nye County, Nevada; Naturita, Colorado; and Edgemont, South Dakota. Facilities ranging from nuclear reactors to uranium mines to weapons plants contributed large amounts of money to state and, especially, local economies. Particularly in rural areas, jobs at nuclear facilities provided incomes which far exceeded those available in other employment. Through its pool of financial resources, the nuclear establishment created and concentrated positive incentives for locations willing to accept its activities. Intangible, vaguely defined, potential environmental impacts from a nuclear facility competed with the promise of tangible, quantified, positive, and immediate economic rewards. Thus, the nuclear establishment built pockets of geographically concentrated political support for its activities, and its public relations and education efforts promoted the benefits of nuclear facilities.

Conflict over how to define the problem of high-level nuclear waste disposal implied a set of spatial considerations and questions. Consider the geographical extent of the problem. Was nuclear waste a national problem, a regional one, or one that should be left to the individual utilities? What states, regions, jurisdictions, and constituencies should participate in decision making? In what forums should the conflicts be resolved? What types of institutions should be responsible for selecting a site and implementing the solution? With its resources and access to national political institutions, the nuclear establishment was well placed to promote its own answers to these questions. High-level waste disposal could be defined as a national problem that demanded federal involvement. Their power could also be used to promote location-specific issues and constituencies. If localized impacts became a problem, compensation could be offered. Locational issues moved debate away from more fundamental questions: should nuclear waste generators (that is, reactors) be closed or phased out? Should there be a national referendum on the continued use of nuclear power (for example, as was held in Sweden and Austria)? Should a new set of institutions for radioactive waste management be established? Were new forms of scientific peer review needed to assess waste management plans and technologies? Was it legitimate to have DOE—the promoter of nuclear power—administer a civilian waste disposal program? In short, the political-geographic definitions of issues promoted by the nuclear establishment left no room for fundamental questions about legitimacy, credibility, or social priorities.

Because the atomic energy acts established federal authority to regulate the nuclear industry and exempted military facilities from most regulations, state and local entities had a difficult time forcing the nuclear establishment to recognize the legitimacy of their interests and concerns. Environmental, public interest, scientific, and other groups from civil society promoted greater control and oversight over the nuclear establishment. In these disputes members of the nuclear establishment lined up to oppose the expansion of nonfederal authority and additional public intervention. Attempts to expand public oversight were countered with the argument that parochial interests would only use this additional layer of burdensome, inefficient regulation to destroy an industry crucial to national security and the economy. In some cases, however, state and local entities were recognized for their interests and given authority to restrict the nuclear industry's activities. Notably, the Supreme Court allowed states to use their rate-setting authority to restrict the construction of new nuclear power plants. In that case, states argued that the lack of a demonstrated disposal technology for high-level waste created economic uncertainties which made it impossible to set fair rates for electricity based on utilities' costs.[19] Nevertheless, political challenges from state and local institutions and allied civic groups continued to be blamed for obstructing progress toward a national solution to the waste disposal problem.

CHAPTER 3

DISORDER AND CRISIS

FORTY YEARS after the first nuclear chain reaction was produced at the University of Chicago, Congress passed the first piece of legislation to explicitly address the management of high-level nuclear waste. Until then, nuclear waste management had been an administrative program conducted at the discretion of the AEC and, later, the DOE. Lack of a specific nuclear waste policy act did not prevent either agency from conducting site screening surveys, research on disposal technologies, and site development work (see table 2). But after efforts in the early 1970s to site a repository at Lyons, Kansas, failed—when state geologists revealed serious problems with the site—the issue of how to handle high-level waste received more attention. Still, federal legislation was not forthcoming. However, by the end of the decade utilities were aggressively pushing for federal legislation. The Sierra Club declared radioactive waste one of its top three issues for 1980. In that year thousands of pages of congressional testimony on high-level waste disposal were collected and congressmen such as Morris Udall, introduced comprehensive nuclear waste legislation. In 1982 a spokesman for the utility industry declared that "the need for nuclear waste legislation cannot be denied." What changes precipitated demands for a legislated solution rather than continuing on with DOE's program?

By the end of the 1970s, various social, economic and

TABLE 2

MILESTONES IN THE U.S. HIGH-LEVEL NUCLEAR WASTE PROGRAM PRIOR TO
PASSAGE OF THE 1982 NUCLEAR WASTE POLICY ACT

1957	National Academy of Sciences panel recommends salt deposits as a possible repository site
1963–1967	Tests conducted in salt deposits at Lyons, Kansas
1968	Evaluation of sites in the basalt formation at Hanford, Washington, begins
1970	Pilot repository in a salt formation at Lyons, Kansas is proposed
1972	Lyons, Kansas, site withdrawn; search shifted to salt formations in New Mexico
1974	Waste Isolation Pilot Plant (WIPP) facility proposed for New Mexico
1976	National Waste Terminal Storage program announced
1978	Office of Nuclear Waste Isolation formed at Battelle Memorial Institute
1979	Interagency Review Group on Radioactive Waste Management publishes its report
	Studies in progress at WIPP, the Nevada Test Site, Hanford, and four salt basins–including Utah and the Gulf Coast
1980	Final environmental impact statement, "Management of Commercially Generated Radioactive Wastes," is issued
1981	DOE program policy modified to focus on intensive site investigations and accelerated construction of a facility
1982	Passage of Nuclear Waste Policy Act permits grandfathering of already-studied sites into a new site selection process headed by DOE

political changes threatened to disrupt relationships among members of the nuclear establishment and the basis of their power. The expertise of nuclear scientists and engineers and the legitimacy of DOE programs were being questioned. The health of the nuclear economy was waning; and nuclear waste legislation was looked to as one means of maintaining order and stability. However, various opponents of the nuclear establishment demanded new government agencies to oversee nuclear facilities, an emphasis on environmental quality over economic growth, stricter regulations and environmental standards, an end to federal promotion of nuclear power, and decision-making processes more responsive to local concerns. In short, they ordered new patterns of relations among the state, economy and civil society. This chapter begins by discussing conditions which produced disorder in the nuclear establishment, followed by an examination of three crises which threatened the nuclear establishment's control over the management of high-level waste.

A major development in U.S. society which challenged the nuclear establishment was the emergence of a view of the priorities and functions of the state which emphasized environmental protection, public health, and safety. Environmental legislation and public awareness of these issues had steadily increased through the 1970s. An extraordinary amount of environmental legislation was passed in the 1970s in spite of a retreat from liberal social policies, an energy crisis, and economic recession. Most important, political action was increasingly focused on the hazards of industrial technologies. The politicization of chemical manufacturing paralleled that of nuclear power. Public perceptions of the risks associated with the manufacture of pesticides and toxic chemicals closely approximated perceptions of the risks of nuclear power and waste disposal.[1] The discovery of abandoned toxic waste landfills paralleled the rediscovery of radioactive waste sites at manufacturing plants in Chicago, Pittsburgh, and Salt Lake City. To the public, the actions of the Hooker Chemical Co. at Love Canal may not have appeared much different from those of nuclear industries. Corporations who manufactured hazardous chemicals—including Du Pont, Union Carbide, and Allied—also played a major role in U.S. nuclear programs. Through the 1970s and into the 1980s public opinion polls showed increasing concern about hazardous waste issues and support for environmental clean-up programs. Public support for nuclear power declined. According to studies sponsored by the utility industry, in 1975 sixty percent of Americans polled favored the construction of more nuclear power plants. By 1983, sixty percent opposed more construction.[2] Moreover, public concern for environmental

protection and health were not the only issues that translated into criticism of the nuclear establishment. Political
opposition to chemical and nuclear industries, particularly
those involved in waste disposal, accompanied the emergence of political ideologies critical of all large-scale, cen-
trally controlled technologies.

Critiques of centralized technologies were assimilated
into the rhetoric and ideology of environmental and local
political activists.[3] Discussions of alternatives to centralized
energy technologies appeared in congressional testimony by
Amory Lovins, whose 1976 article on "soft energy paths"
in *Foreign Affairs* attracted the attention of policy makers
and created a new language of energy policy discussion.[4] Nuclear power, the epitome of centralized, bureaucratized technology, was increasingly used as a symbol of what was
wrong with the American state. In his review of the technological risk literature, Covello found numerous reports supporting the observation that "disputes about nuclear power
are often about values and goals that far transcend issues
of health and safety. Many people are concerned about nuclear power not because of its specific risks but because of
its associations with nuclear weapons, highly centralized
political and economic systems, and technological elit-
ism."[5] Nuclear power and nuclear waste disposal were thus
more than issues of energy policy. They were symptoms of
a fundamental tension within U.S. society between members of the nuclear establishment, who asserted the priorities of the national state and economy, and their opponents,
who would strengthen local interests and community self-
determination.[6] Ideologies of environmentalism, which emphasized local quality of life, confronted the nuclear establishment's emphasis on national security and technological
progress. Opponents of the nuclear establishment saw
DOE's nuclear waste siting program as yet another example
of how the national state's promotion of big science and
big technologies ran roughshod over the state and local governments' mandates to protect public health and safety.

A second source of disorder threatening the nuclear establishment followed the implementation of new forms of political participation intended to permit greater public oversight of bureaucratic decision making and increased access to technological resources. The 1970s saw widespread changes in political participation as public review of state policy making became a procedural requirement. Like other federal agencies, the NRC was required to hold its deliberations in public, keep records of its proceedings, and permit the public to contest agency decisions. The heretofore smooth operation of federal agencies was disrupted by the need to address public comments and submit proposed decisions to public scrutiny. The 1969 National Environmental Policy Act (NEPA) brought government planning and decision making out of the back office. The nuclear establishment was quick to criticize public participation requirements in its call for limits on "redundant" public hearings, "political and regulatory gimmickry," and costly, repetitive judicial review.[7] Grafted onto early waste policy acts were proposals for limiting the application of NEPA to nuclear facilities, restricting judicial review, issuing operating licenses without public hearings, and otherwise limiting public intervention in the operations of the nuclear establishment.

In addition to NEPA, the 1966 Freedom of Information Act created new tools for breaking federal agencies' control over technical and performance data; it also produced "sunshine laws" which opened administrative hearings to the public. In one widely noted application of the Freedom of Information Act, Daniel Ford's *The Cult of the Atom: The Secret Papers of the Atomic Energy Commission* (based on heretofore classified documents) embarrassed the nuclear establishment by documenting how mission-oriented officials of the AEC responded to warnings from their own scientists about unsafe reactors by suppressing the alarming reports and pressuring the authors to keep quiet.[8] Changes in tort law expanded definitions of public risk and responsibility, increased chemical and nuclear industries' exposure to legal action, and even pressed some into bankruptcy.[9]

Local activists successfully placed nuclear waste, liability, and safety issues on referendums in seven states in 1976; three in 1980 and again in 1982. At the national level, the growth of congressional staffs accompanied greater involvement and oversight of administrative programs. Congress became less willing to defer to the judgment of the AEC or DOE experts.

A third source of disorder was the emergence of new forms of power. Along with increased local support for environmental protection and an expanded legal base from which to contest federal actions, state and local governments developed sources of authority and expertise apart from those of the nuclear establishment. Much federal environmental legislation was turned over to the states for implementation. The trend toward state implementation of environmental legislation—initiated with the 1963 and 1965 Clean Air Acts—continued in the implementation of hazardous waste legislation such as the Resource Conservation and Recovery Act (1976) and the 1980 Low-Level Radioactive Waste Policy Act. In those years some states built sophisticated organizations of technical and legal personnel capable of addressing complex environmental issues, and, in many cases, state actions preceded federal legislation or resulted in regulations more stringent than federal standards.[10] The Minnesota Pollution Control Agency, for example, attempted to impose radiological discharge standards more stringent than the AEC's on a reactor located near the Twin Cities. And while a federal court of appeals in a 1971 decision (*Northern States Power Co. v. Minnesota*) sided with utility arguments that federal authority preempted the state's, the case was evidence of states' abilities to mount sophisticated challenges to the nuclear establishment. California eventually went to the Supreme Court where, in 1983, it won a decision upholding the constitutionality of its 1976 statute prohibiting the construction of new reactors until a permanent solution for the high-level waste disposal problem could be demonstrated. Other political action groups also developed the experience and expertise to sponsor effective challenges. The Natural Resources Defense Council, as one example, was in court in the mid-1970s trying to impose civilian environmental standards on DOE nuclear facilities. Its efforts brought public attention to environmental violations at DOE weapons plants.

Members of state legislatures, city councils, and other local authorities sponsored and often passed ordinances and regulations on power plant siting, nuclear waste disposal, and transport.[11] However, aggressive local regulation threatened to destroy two major advantages of federal regulation: (1) federal regulation established a regulatory uniformity desirable for economic institutions to conduct interstate business with a minimum of local interference; and (2) it supported economies of scale dependent upon the adoption of standardized business and environmental practices. The nuclear establishment thus demanded federal legislation which would restore order and eliminate the proliferation of renegade state and local regulations that complicated its operations.

By 1980 these three changes in U.S. politics and society formed the context within which the disorder of the nuclear establishment reached crisis proportions. In 1971 the AEC predicted that fifteen hundred gigawatts of nuclear capacity would be on line by the year 2000; by 1980, those projections had been scaled back to a few hundred (and were reduced again to less than 125 in 1984). A total of eighty-seven proposed and in-progress nuclear plants were abandoned between 1975 and 1983. By 1982 the bill for abandoned nuclear plants totaled ten billion dollars.[12] Construction time for the Diablo Canyon nuclear power plant exceeded sixteen years as local activists staged acts of civil disobedience and challenged the facility's seismic integrity in hearings and lawsuits. Protests occurred at Seabrook and numerous other reactor sites. The final cost of Long Island Lighting Company's Shoreham nuclear power plant was at least fifteen times the original estimate. By 1981 the possibility of the Washington Public Power Supply system defaulting on billions of dollars in bonds to finance a nuclear power plant was evident. Responding to demands for improved reactor safety, between 1970 and 1980 over three hundred new AEC-NRC regulatory guidelines were issued. Retrofit of reactors and design alterations consumed billions of dollars. Problems with the disposal of low-level waste also reached critical proportions when governors with waste sites threatened a shutdown unless new sites were developed. Some utility spokesmen claimed their opponents were using waste disposal and transportation issues to campaign against nuclear power rather than confronting "real issues" such as "the impacts that inadequate electrical generation capacity might cause." They called for complete federal preemption of all decisions related to nuclear technologies and called public review groups and consultation with the states a waste of time.[13]

Within this context of change, challenge, and disorder, the nuclear establishment faced three crises: (1) a crisis in the credibility of its decision-making procedures and expertise; (2) a crisis in the perceived legitimacy of its institutional authority; and, (3) a crisis in its financial viability. The nuclear establishment looked to a nuclear waste policy act for rescue.

The First Crisis: The Loss of Scientific Credibility

The AEC can be described as a "New Deal type of agency" whose legacy carried over to its successor, DOE.[14] Briefly, the New Deal model of a bureaucracy was one in which professional, similarly trained, expert decision makers were insulated from public and political pressures so that they could make optimal decisions on the basis of objective scientific criteria. The assumption was that rational methods of planning and decision making produced efficient, scientifically valid, and defensible conclusions. Social desirability was equated with engineering efficiency. The expert planner was handed control of decision-making criteria, procedures, and assessments of risks, costs, and benefits.

In his discussion of bureaucratic power and expertise, Francis Rourke asserts:

> In all modern societies, whether democratic or nondemocratic, a first and fundamental source of power for bureaucratic organization is the expertise they command—the greatly varied skills that administrators bring to the policy process.
>
> . . . .
>
> [Quoting Weber] "Under normal conditions, the power position of a fully developed bureaucracy is always overpowering. The 'political master' finds himself in a position of the 'dilettante' who stands opposite the 'expert,' facing the trained official who stands within the management of administration."[15]

A bureaucracy takes complex problems, breaks them into components, and assigns them to specialized units within the organization. Specialized knowledge of these particular problem areas makes the expert the master of the situation, especially when the norms and topics of scientific investigation are subject to the bureaucracy's control or the facts cannot be subjected to outside verification. Monopolistic control of the facts reinforces the power of bureaucracy. Policy makers' reliance on expert advice leaves the latter in a position to shape value and factual premises and influence the definitions of problems and their solutions. In this way the assumptions underlying nuclear waste policy were shaped by the bureaucracy of experts.[16]

A credibility crisis occurs when these expert decisions and decision-making processes are challenged and their limitations or failures are revealed. In the case of nuclear waste, expert advice and the certainty of a solution became suspect as barriers to specialized knowledge were broken down and alternative facts were produced. Broad sectors of society were no longer willing to simply defer to the judgment of the establishment's scientists and engineers. Such a crisis threatened the nuclear establishment's control over the language and content of policy making. Certainty and control, which Beneviste sees as two key sources of the power and attractiveness of expertise, were replaced by uncertainty and disorder.[17]

The early years of nuclear power were filled with statements about the certainty of science's ability to solve social problems (by providing cheap energy) as well as any technical problems that might arise during the commercialization of nuclear power. Confidence was so great that pilot projects for testing and evaluating the first commercial reactor designs were largely dispensed with. Reassurances that the waste produced by reactors was a nonproblem were not simply the hyperbole of the industry's promoters. In 1957 the National Academy of Sciences (NAS) stated that salt appeared to be "the most promising method of disposal of high level waste" and that "the committee [NAS Advisory Committee on Waste Disposal] is convinced that radioactive waste can be disposed of safely in a variety of ways and in a large number of sites in the United States."[18] The NAS report instilled confident complacency. A 1960 nuclear power plant siting survey in New York found that fifty-seven percent of the people interviewed were confident in the AEC's ability to dispose of nuclear waste. Another thirty percent had no opinion and thirteen percent expressed some lack of confidence. However, in 1973, after the liquid high-level waste leak at Hanford, Washington, was widely publicized by the media and environmental groups, various surveys found that nuclear waste was beginning to be viewed as a serious problem associated with nuclear power. Opponents of nuclear power latched on to the waste management issue.[19]

The nuclear establishment used the NAS report, which was widely quoted for decades afterward, to defend its assertion that engineered solutions would be readily available when needed. During years of hearings on nuclear waste policy acts one commonly heard the argument that "the technology was in hand." The utilities often referred to yet another NAS report, *Energy in Transition: 1985-2010*, which stated: "No insurmountable technical obstacles are foreseen to preclude the safe disposal of nuclear wastes in geologic formations. All necessary process steps for immobilizing high and low-level wastes have been developed, and there are no technical barriers to their implementation."[20] Nuclear waste disposal was not a technical problem but a problem produced by burdensome regulation, political delay, and inaction, the implication being that, if the politicians and political groups stayed out of the way, the experts could solve the problem. A speaker for the Atomic Industrial Forum observed: "the risks from such a repository are below the level that we are generally aware of or concerned about in going about our activities after getting out of bed in the morning. ... We are spending our time inventing non-technical institutional impediments to the solution. ... [and] institutional impediments of questionable value"[21] Despite such confidence, the disposal of high-level waste proved to be a very difficult problem. The Hanford tanks, projected to not leak for thirty years, leaked in ten. At Lyons, Kansas, the AEC identified an old salt mine as the best site for a high-level waste repository. This choice, however, was discredited by state geologists who found evidence of previous drilling and potential leaks in the salt formation, which was overlooked by federal experts. Subsequently, a variety of other programs, sites, and technologies were tried. Attempts to explore potential repository sites in Michigan without notifying state or local officials provoked vocal opposition. The Energy Research and Development Administration said it was only conducting preliminary exploration work, while Union Carbide, who actually ran the national waste terminal storage program for ERDA, said that it was in the final stages of selecting a site. In 1976 a congressional hearing was held in Alpena, Michigan, during which the Hanford leaks and Lyons case were used as evidence of waste mismanagement, the AEC's and ERDA's "arrogance" and technical incompetence, and misplaced confidence in risk assessments. State and local officials demanded the hiring of outsiders to review ERDA's technical data.[22] Similar conflicts erupted in Ohio and that area was dropped from

consideration. The EPA rejected an ERDA environmental assessment of an alternative to a geologic repository—a retrievable surface storage facility which would not depend
upon locally specific geologic conditions and thus could
be sited virtually anywhere. Proposals were also made for
shooting the waste into the sun, dropping hot canisters
where they could burn their way into the polar ice cap, or
jettisoning them into oceanic trenches. By the late 1970s
the nuclear establishment's failure to design and implement
a safe, publicly acceptable waste disposal solution had been
repeated many times. Edward Wiggin of the Atomic Industrial Forum called the lack of permanent waste disposal "the
biggest public acceptance problem" for the nuclear industry.[23]

Repeated mistakes and failures seriously compromised
future attempts to address the waste disposal problem since
they diminished public confidence in the nuclear establishment's experts. A growing number of scientists—such as
those who left General Electric to join the Union of Concerned Scientists—lent their expertise to dissident groups
critical of nuclear science. Critics achieved notable success
in producing alternative sources of data which challenged
the risk analyses of the nuclear establishment. The Rasmus-
sen Reactor Safety Study (Wash-1400), which "proved" the
safety of nuclear power and dismissed the environmental
effects of accidental radiation releases, was withdrawn after
intense criticism. The scientific credibility of the nuclear
establishment was shown to be vulnerable and at times blatantly self-serving. In the case of nuclear waste, Morris
Udall summarized this erosion of confidence:

> You [Prof. Paul Witherspoon, University of California-
> Berkeley] leave me with kind of a pessimistic feeling here. I first
> got into this task in the 1960s, and it became a cliché in the
> 1960s and 1970s to say that there was no engineering or scien-
> tific problem with nuclear waste, it was simply a political
> problem.
>
> It seems today, in light of your testimony and some of the
> other things we have heard, that we may have a hell of a lot

of engineering and scientific problems that aren't resolved yet from the standpoint of coming up with a [waste disposal] program.[24]

Growing evidence of power plant accidents and waste mismanagement reinforced the loss of credibility. Access to data which could be used to prove charges of mismanagement and technical failure increased when the NRC made public its investigations of reactor safety violations. Reports of nuclear accidents became commonplace in the mass media; bizarre accidents attracted public attention. In 1975 a fire at the Browns Ferry, Alabama, reactor, started by a technician checking for leaks with a candle, knocked out fifteen percent of the electrical capacity on the TVA grid. The story was published in *Newsweek* magazine, and stories of "nuclear slapstick" became common.[25] It became difficult to sustain a confidence in engineered solutions under such assaults. Nineteen seventy-nine was a very bad year for the nuclear establishment with Three Mile Island, the Kerr-McKee/Karen Silkwood verdict, the release of the movie *China Syndrome*, and an antinuclear march on Washington attended by tens of thousands of people. A history of covering up poor engineering and safety violations, suppressing evidence on radiological risks, and falsifying test results had compromised the purported commitment of the nuclear establishment to conservative engineering and safe operations.

Scientists have puzzled about the difference between nuclear experts' and the public's assessment of the risks associated with nuclear reactors and waste disposal. The difference says much about each group's confidence in political-scientific institutions and public confidence in the expertise of the nuclear establishment. For example, in 1982, a Cambridge Reports study found that twelve percent of the population regarded the testimony of utilities on energy matters "very believable" and only eight percent gave that rating to the nuclear industry; in the same study twenty-four percent voted Ralph Nader "very believable," and thirty-seven percent gave scientists from a leading university the same rating. Differences in risk perceptions reflected conflict among sources of expert authority, leadership in science, and scientific credibility.[26] In general, scientists and engineers within the nuclear establishment adhered to an individual-rationalist view of science as a process for converging on the truth which must be free from political-economic pressures and social biases. In contrast, their opponents maintained that political-economic biases were already built into the science and analyses produced by the nuclear establishment.

It is no surprise, then, that public distrust of nuclear establishment science and expertise increased when predictions about the reliability and cost-effectiveness of nuclear power did not materialize or when consumers saw their utility bills increase as nuclear power plants were brought on line or abandoned. Given this loss of credibility, the growth of activism in the form of antiutility, antinuke, anti-DOE, antireactor groups was not a surprising development.[27] Many nuclear failures and embarrassments involved technologies other than waste disposal, such as the controversy surrounding the AEC's suppression of data about the harmful effects of nuclear testing in Nevada. In the end, such incidents further compromised the credibility of the entire nuclear establishment. This loss of belief, reinforced by a broader social trend toward distrust of large-scale technologies, meant that virtually every recommendation and scientific study produced by the nuclear establishment would confront intense public skepticism, scrutiny, and distrust. Convergence on a scientifically valid solution would first have to overcome this loss of credibility.

The Second Crisis: The Loss of Institutional Legitimacy

In his study of what is really the nation's first (albeit a military), waste repository—the Waste Isolation Pilot Plant (WIPP) near Carlsbad, New Mexico—Gary Downey found, "The key issue in the dispute concerned the *political legi-*

timacy of decision-making mechanisms for repository siting, which depend upon the extent to which they both adequately represent the interests of affected groups and meet an indistinct technical/political criterion of acceptable safety" (emphasis added).[28] Legitimacy is key to the exercise of bureaucratic authority within a democratic system. The legitimacy crisis originated in challenges to the rules and procedures of decision making and the policy-making process itself. The legitimacy of institutional processes and procedures can be maintained by the traditions that permeate a society whose acceptance inhibits even considering the possibility of challenging the existing order or creating an alternative political process. This source of legitimacy is rooted in deference to authority. In this case, a crisis occurs when the possibility of successful challenge becomes apparent, when claims of traditional authority and procedures are questioned, and when opportunities are seized to politicize an issue.

A second basis of legitimacy may be termed rational-legal as it refers to principles of law and administrative procedure and the network of rights, duties, and authority attached to formal offices. By invoking principles of law or procedure, an institution may claim that it is the only entity with the authority (and expert knowledge required) to address certain problems or issues. Such claims are always tenuous and can be changed by judicial and legislative action. In this case, the crisis demands new political institutions and processes, a redistribution of privileges, and counterclaims to legal authority.

A third source of legitimacy originates in charismatic promises to relieve social distress, guarantee economic and social well-being, and to achieve goals of mythical proportions—such as energy too cheap to meter. The failure of these dreams to materialize or the attractiveness of competing dreams undermines these claims to legitimacy.[29] The failure to realize utopian dreams may be exacerbated by counterdemonstrations of alternative means for achieving the same benefits—such as lower energy costs through conservation rather than nuclear power plants. The crises of legitimation may stem from political movements which interject new values into political life. As people are drawn to new ideals of environmental quality, world peace, or a nuclear-free society, this crisis becomes a reaction to the "scientization of politics"—an alienation from politics which has been reduced to arcane technical questions and expert decision making. In this sense, conflict consists of more than competing political interests. It points to the inability of existing institutions to respond and adapt to changes in their environment.[30]

Even if opponents of the nuclear establishment had insufficient power to replace it with an alternative set of institutions, they could still demand concessions before accepting the legitimacy of a new site selection program. Similarly at times some members of the nuclear establishment conditioned their endorsement of the DOE waste management program. For example, the utility industry sometimes pointed to DOE inaction and mismanagement as the cause of the waste disposal problem and demanded the immediate construction of a pilot waste disposal facility as the key to restoring order.

For members of the nuclear establishment, the management of spent fuel and high-level nuclear waste disposal was defined as a uniquely federal responsibility. In 1977 an ERDA spokesman said that, through the history of commercial nuclear power, virtually everyone had assumed that all aspects of the nuclear fuel cycle—*with the exception of permanent high-level waste disposal*—would be created by industry action. Morris Udall declared, "We [Congress] owe the country a decision to take the stuff and get it stored the best way we know how."[31] A spokesman for the NRC, testifying before a 1977 congressional subcommittee, said that the federal commitment to accepting utilities' high-level waste began with the Atomic Energy Act of 1954. ERDA declared it had "the responsibility to accept and dispose of what's declared as high-level waste."[32] Nuclear utilities claimed that the federal government assumed responsibility for the disposal of spent fuel when President Carter decided to defer reprocessing in the interest of nuclear non-proliferation. In the words of a spokesman for the Edison Electric Institute, "The federal government decision to defer or forego commercial reprocessing requires, at a minimum, a firm federal government commitment to acceptance of spent fuel for shipment to a government repository as soon after discharge as practicable."[33] According to the utility industry, since the federal government created the problem, the federal government should solve it. In October 1977, the DOE announced its intention to eventually take legal title to all spent fuel held by utilities.

Lacking a specific congressional direction for nuclear waste management, the DOE relied on the Atomic Energy Acts and old AEC regulations as justification and precedent for its authority to define a high-level waste program. It was up to the DOE to decide how this program would be implemented and at what points other groups would be involved. However, considering the repeated failures and abuses of past administrative programs, Congress appeared increasingly reluctant to grant DOE continued independence. Supporters of nuclear power viewed past failures to solve the nuclear waste problem as fueling public perceptions of incompetence and lack of resolve. Senator Charles Percy, declaring that the nuclear industry saw the DOE as "irresolute, unwilling to take a stand on nuclear waste," introduced legislation which said that "the federal government should assert responsibility for long-term care of all wastes." Percy's legislation also aimed to reassert DOE as the primary waste management agency and uphold Congress' traditional role of providing policy leadership and shaping a national consensus on the solution to this problem.[34] Confronting the "erosion of public confidence" in the federal government's ability to resolve the nuclear waste issue became a major concern of Congress.[35] Thus, various accelerated site selection processes and fast-track demostration projects were proposed to establish federal authority, commitment, and resolve.

Disorder was accelerated by lurching shifts in federal policy and management. Utilities, in anticipation of spent fuel reprocessing, had sized their storage pools accordingly and were thrown in turmoil when the reprocessing option was abandoned. Reprocessing was restored by President Reagan too late. Federal agencies first claimed that interim storage of spent fuel—until a repository was on line—was a federal responsibility, then a private, and eventually settled on defining it as a joint responsibility. Initially, spent fuel was not considered as waste but would be stored to allow retrieval and eventual recovery of its uranium and plutonium; later, it was defined as waste for permanent disposal. WIPP would be available for commercial spent fuel, then it was not. A repository would be available by 1985, then by 1988, and then not before 1998.

Reorganization of bureaucracies and procedural instability exacerbated the turmoil created by shifting technical and management priorities. With the Energy Reorganization Act of 1974 the AEC was split into the NRC and ERDA. Within two years of its establishment ERDA merged with the Federal Energy Administration to become DOE; four years later, President Reagan threatened DOE with a merger into the Commerce Department. The Department of Energy's internal reorganizations disrupted program continuity and produced procedural instability. Site selection studies, criteria, procedures, schedules, and priorities changed at the discretion of DOE administrators, making it difficult for outsiders to identify the specific plans and decision points leading toward development of the repository. In 1978, Senator McClure noted that these program changes and uncertainties made it difficult to tell who actually had control of waste management.[36] The Natural Resources Defense Council attacked DOE's lack of a clear decision-making process and site screening criteria, its lack of a "rational process of careful problem definition," and its pursuit of site selection without any knowledge of possible EPA environmental and performance standards for a repository.[37] As the NWPA was being debated in 1982, states and environmental groups charged the DOE with a history of procedural inconsistencies, such as announcing decision-making processes and schedules and then repeatedly modifying them. The Department of Energy claimed it was committed to working with the states and then entered areas unannounced to conduct exploration drilling. They issued reports for public comment after decisions had been made, and agreed to a full EIS review and then later opposed it. DOE announced a schedule of site studies and then later, in the Reagan administration, advocated increased funding to bypass studies and fast-track repository development. While agreeing that consumers of nuclear-generated electricity should pay for the repository program, the DOE later supported legislation that would require a host state to pay twenty-five percent of the project's impact mitigation costs. At another point, the DOE said that the federal government should assume all such costs. Although they had advocated in-depth site characterization studies, the DOE later tried to eliminate most studies and accelerate construction authorization and licensing of a repository—announcing in June 1982 that shaft construction at the Hanford and Nevada test sites would begin in 1983.[38] By 1982, changing program priorities and elements had eroded confidence in past DOE commitments and efforts to gain state and local cooperation. James Tierney, Maine's attorney general, observed in testimony before Congress, "The people just do not trust the process itself, the procedures, or the governmental agencies involved, and you are going to have to be sure that any process coming out of the Congress in dealing with nuclear waste is fair and open to everyone."[39] State governments demanded that federal legislation set decision-making procedures, criteria, and environmental reviews and require full application of NEPA. A Sierra Club spokesman observed:

> Given the past record of federal efforts in managing nuclear waste, it is naive to expect the public to accept the uncertainties

that are inevitably present in any proposed "solution" to the waste problem unless it is clear that the *institutional process* will provide an accurate and straightforward test of the safety and validity of the program. To the extent that proposed legislation attempts to short-circuit key environmental and licensing reviews, both the public's confidence and the *integrity of the process* itself will inevitably be eroded (emphasis added).[40]

At the same hearing, the Office of Technology Assessment asserted that procedural requirements should take precedence over meeting industry desired schedules and deadlines.[41]

As federal agencies vacillated on what to do with spent fuel, state and local agencies pushed at the boundaries of their authority. By 1979, nearly thirty states had enacted legislation to regulate spent fuel storage, despite claims by members of the nuclear establishment that most of this legislation was unconstitutional. Statutes similar to California's—which barred new nuclear power plant construction until a disposal solution could be demonstrated (declared constitutional in 1983)—were adopted by Massachusetts and Maine. Illinois banned the storage of spent fuel originating from outside the state; Iowa banned construction of nuclear plants until a reprocessing technology was available; Minnesota prohibited a waste management facility in the state, as did Ohio, Louisiana, and others.[42] Numerous state and local regulations on waste transport were also enacted, and more regulations would follow in the 1980s. In the Reagan administration, lackadaisical enforcement of environmental regulations and the botched management of the Superfund program contributed to the perception of federal disinterest in public health and environmental issues. By all appearances, the administration's renewed promotion of nuclear power compromised federal attempts to regulate the nuclear industry in the interest of protecting the public and the environment. Thus, many state and local governments moved into the vacuum created by federal neglect and indifference.

The Third Crisis: Financial

The legitimacy crisis reinforced challenges to continued federal support for commercial nuclear energy and the reliance on DOE for implementing a waste disposal program. The crisis in scientific credibility produced questions about the dependability and economic advantages of nuclear power. The third crisis producing disorder and uncertainty in the nuclear establishment was financial. In the context of widespread restructuring and stresses in the U.S. and world economy, the nuclear establishment confronted at least two serious political-economic developments: (1) declining support for state-subsidized energy projects, and (2) the declining profitability of U.S. nuclear utilities and other nuclear industries. Because the financial problems of the nuclear establishment have been well documented, the remainder of this chapter will only touch upon the highlights.[43]

In the 1970s, slowed economic growth, recessions, and the adoption of energy conservation practices reduced the need for new power plants. Whereas sales of electricity grew over six percent per year in the 1960s, in the 1974 recession the growth rate plunged to zero and showed negative growth in 1981–82.[44] Patterns of energy consumption also changed. Rising prices encouraged energy intensive industries to invest in new technologies and management strategies which achieved significant reductions in their consumption of electricity. Energy conservation became a growth industry as consumers invested in new technologies, energy efficient appliances, and changed household practices. Large industrial consumers of electricity, such as smelters, faced hard times and shut down or cut back on production, creating surplus generating capacity. Many nuclear power plants whose construction began in a time of accelerating demand would be coming on line in a period of flat or slow growth in demand. Investor confidence in electric utility stocks, heretofore a conservative investment, plunged along with stockholder equity. In 1970, Standard and Poor's rated the debt of seventy-nine utilities as AA or better. In 1980, thirty

made the cut. Utilities with nuclear power plants showed
a lower rate of return than those without them. Following
the accident at Three Mile Island, a Merrill Lynch study
found that a utility's use of nuclear power was considered
a risk factor by conservative institutional investors. Solomon Brothers advised investors to be wary of nuclear utilities.[45] Once the safest of all investments, the electric utility
industry was shocked by threatened bankruptcies and de-
faults on bonds.

Utilities, one of the most capital intensive industries,
faced OPEC-like increases in borrowing costs in the latter
half of the 1970s. The final cost of new plant construction
and licensing ballooned. The nuclear industry blamed
changes in regulations and burdensome NRC procedures for
rising costs. Later analyses published in the financial press
pointed to mismanagement of plant construction and design
as a prime factor.[46] Increasingly sympathetic to charges that
mismanagement was responsible for rising costs, state utility commissions became more and more reluctant to grant
utilities the rate increases needed to recoup the full cost
of constructing and operating nuclear power plants. In cases
where cost overruns were due to mismanagement, the commissions often required that they be assumed by the utility's
stockholders. The structure of the electric utility industry
itself was changing. The Public Utility Regulatory Policy
Act of 1978 required utilities to buy electricity from nontra-
ditional sources. Cogeneration emerged as a lucrative new
energy field that could be entered by virtually any industry—from sawmills and paper plants to petrochemical
plants. Engineering companies that had lost business due
to cancellations of nuclear power plants entered this new
field. Industrial consumers broke off contracts to buy a percentage of new power output and considered installing their
own electrical generating equipment and competing directly
with the utilities.

All through this financial crisis the utility industry continued to face two problems: where to temporarily store the
spent fuel already filling reactor pools and what, in the long

term, to do with it. The Carter administration decided that interim, temporary storage of spent fuel should be left to the utilities until a permanent federal disposal facility was operational. However, the DOE and the nuclear industry claimed that political and regulatory uncertainties meant that "private industry [i.e., suppliers of AFR—away from reactor storage—or nuclear utilities] has serious difficulties facing the business risk or uncertainty of providing new AFR storage facilities. . . . The industry cannot forecast costs associated with facilities and facility operation, nor the projected amount of storage required, nor the time period for which storage would be required under the existing regulatory climate."[47] Unlike a utility-owned facility, the DOE stood a better chance of circumventing NEPA, NRC licensing, state regulations, and state public utility commissions—avoiding many costly conflicts and delays. The DOE estimated that conversion of inoperable, spent fuel reprocessing plants into temporary storage facilities would recover $362 million Allied had invested in Barnwell, South Carolina. It would also recover General Electric's investment in its Morris, Illinois, facility, which was already being used to store some spent fuel rods. Opponents viewed proposals for a federally owned temporary storage facility as yet another subsidy for nuclear power. Congressional critics, such as Representative Edward Markey of Massachusetts, were quick to claim that DOE's 1977 decision to take title to utilities' spent fuel encouraged General Electric to abandon plans to enter the temporary storage business and utilities' unwillingness to invest in reactor site storage.[48]

Utilities, which never planned to be responsible for the final disposition of high-level waste, were relieved when DOE contracted to take title to spent commercial fuel by 1998. However, they still faced the issue of who would pay for the repository. Payment out of the federal budget was unlikely during a time of campaign calls to trim spending and reduce the deficit. Apparently, industry or consumers would have to cover the cost of waste disposal, but the last thing the utilities wanted was another attack on its earnings and stockholders. The key question was whether the cost could be passed on to the ratepayer. A one mil per kilowatt-hour tax on nuclear generated electricity could solve the problem. Under state PUC regulations, a federal tax was a legitimate business expense which could be passed on to customers. In addition, the tax was advantageous because it appeared equitable and, because it was so diffusely distributed, the tax burden was not likely to generate ratepayer opposition. For the nuclear establishment a tax would also free the waste management program from the vagaries of federal funding. Nevertheless, the nuclear industry still advocated shortcuts to reduce project costs, arguing that characterizing more than one potential repository site was unnecessary.

Apart from costs of facility development there was also a price to be paid for winning approval for a repository site. Such costs could take two forms: incentive payments to "sweeten the pot" and convince locals to accept a repository; and compensation payments for actual impacts associated with construction and operation of a repository. In the first case, the American Institute of Chemical Engineers suggested that the location of a repository could be made financially attractive "at a very minor and certainly tolerable cost to the users of electricity." To reduce local opposition some proposed giving the local jurisdiction unqualified grants of millions of dollars per year.[49] Utilities generally supported the idea of compensation to local jurisdictions, viewing it as a necessity to win local approval, and hoping that it would constitute an insufficient percentage of the project budget. However, to reduce such costs, they proposed at one point that states fund twenty-five percent of the bill for local compensation—a suggestion obviously and immediately opposed by states with potentially acceptable sites.

Initially, the Department of Energy assumed that repository development would produce a local economic boom sufficient enough to compensate local jurisdictions for suffering any negative impacts. Host states were not convinced that the promised prosperity would materialize. Repository construction would employ thousands, perhaps tens of thousands, but repository operations could be handled by a few hundred workers. Host states foresaw impacts similar to those suffered by notorious boom towns such as Rock Springs and Gillette, Wyoming. They wanted mitigation strategies to compensate for the infrastructure improvements needed to support a repository and to lessen the impacts of the boom-bust cycle. States such as Utah emphasized this as one of their primary concerns with any repository program.

Crisis and Demands for Change

By 1980 many Americans were showing unprecedented support for policies of environmental protection, public health, and safety, along with increasing skepticism about nuclear technologies. The implementation of new forms of public participation opened administrative decision making to review and criticism. State and local jurisdictions had acquired more sophisticated legal and technical expertise capable of challenging the nuclear establishment. Thus, three changes in U.S. society—new social priorities, forms of political participation, and sources of state and local power—favored the development of major crises which threatened the continuity of the nuclear establishment.

The monopoly on nuclear expertise cracked when opponents of the nuclear establishment produced their own reputable science to contest the scientific credibility and technical reliability of existing waste management programs. Criticism of DOE's organization, site selection procedures, and unresponsiveness to public interests became widespread. A commonly asked question was: how could an agency such as DOE be responsible for the promotion of nuclear energy and at the same time be trusted to critically examine nuclear waste technologies and take a conservative approach toward protecting public health and safety? Historical evidence showed how attention to public health and safety suffered in the enthusiasm to "go nuclear."[50] Finally, the cost of nuclear power was coming under intense scrutiny by state regulators and energy planners. Facing immense capital and financial problems, nuclear utilities and industries were reluctant to take on yet another risky, capital-intensive project–that of interim or permanent waste disposal.

Members of the nuclear establishment came to demand that federal legislation resolve these crises, restore a predictable course to federal waste management programs, and bring order and stability to the nuclear economy. National legislative action was required because it alone could resolve the fundamental conflict between the nuclear establishment's demands for reassertion of its authority and control, and opponents' demands for new institutions and patterns of relations among the state, economy, and civil society.

CHAPTER 4

▼

REASSERTION OR
RECONSTRUCTION

THE THREE crises which threatened the nuclear establishment were not events but developments rooted in the widespread political and economic changes occurring in U.S. society. Similarly, proposals for restoring stability and resolving the conflicts over nuclear waste disposal paralleled the development of the three crises. Legislation which addressed radioactive waste management was introduced into Congress at least six years before passage of the 1982 Nuclear Waste Policy Act (NWPA). In many cases it is difficult to classify a piece of legislation as belonging to the advocates of either reassertion or revision. There was "damn the states, lets get on with it" legislation which blatantly championed nuclear power and the nuclear establishment. And while other proposals sought extensive reconstruction, most of the legislation suggested only minor reforms. A legislative history of the 1982 NWPA could discuss the details of hundreds of proposed nuclear waste policy acts and amendments. Instead, the legislation was examined for what it said about the legitimacy of institutional priorities, power, and responses to political-economic demands; the credibility of past administrative decisions; and the relative importance of economic versus environmental goals and priorities. Proposals which were never incorporated into proposed legislation are important for what they say about unanswered demands for changes in institutional organiza-

71

tion and authority, and for what they say about the likelihood of future conflicts centered on issues of legitimacy,
credibility, and social priorities. For example, abandonment
of nuclear power and the dismantling of the nuclear establishment was never really considered by congressional committees. Thus, keeping such proposals off the legislative
agenda was the first step in the nuclear establishment's reassertion.

Proposals dealing with high-level nuclear waste took
many forms and were produced by organizations ranging
from the Friends of the Earth to the Atomic Industrial
Forum. Table 3 presents a summary of key points made in
those proposals. Utility lobbying groups supported legislation overriding the regulations of the Nuclear Regulatory
Commission (NRC) and the Environmental Protection
Agency (EPA) and limiting judicial review. According to one
utility spokesman, fast-track implementation of a waste disposal demonstration project was critical because, "In my
opinion, [this] waste management legislation is one of about
three requirements for a resumption of nuclear plant orders.
The other two are stabilization of the regulatory process and
restoration of the utilities to financial health."[1]

Investments in Science, Technology, and Credibility

Under Lewis Strauss, the AEC's goal was to commercialize
nuclear technologies and facilitate expansion of the nuclear
industry—two goals maintained by DOE but threatened by
the erosion of its scientific credibility. National political action could help to restore credibility: through legislative
fiat, Congress could endorse the safety and reliability of a
technology, validate past research and development, and remove barriers to the implementation of a waste management technology. A legislation of reassertion could endorse
prevailing definitions of the best available technology, as defined by experts resident in the nuclear establishment. On
the other hand, critics of the nuclear establishment proposed creating alternative scientific institutions unhindered

TABLE 3

SUMMARY OF COMPETING PROPOSALS FOR A NUCLEAR WASTE POLICY

Proposals for Reconstruction	*Proposals for Reassertion*
Give a state veto over site selection; require state participation and consensus	Federal preemption of state regulations and authority; no state veto
Waste problem demands a halt to the construction of new nuclear power plants	No restrictions; construction of additional power plants will be encouraged by a repository
Additional, independent scientific study of sites	Rely on existing knowledge
Additional, independent research on disposal technologies	Use existing technologies and expertise
Include funding of public involvement and citizen participation	Include no funding for participation by citizen groups
Initiate a new national site evaluation process by an independent group or agency	Grandfather sites already studied
Incorporate the principle of regional equity	Emphasize the efficient, least-cost solution
Information-driven process, no hard deadlines, set environmental protection as the primary goal	Schedule-driven process, acceptance of spent fuel from utilities as soon as possible
Create a new lead agency to administer waste management	DOE remains the lead agency
External, independent, intergovernmental review group to monitor the repository and waste management program	DOE answers only to Congress and the president

Implement a process of cooperation and consensus or arbitration with state-local governments	Consultation and information sharing; no arbitration; final decision remains with DOE
Make the waste management agency answer to EPA, NRC, NEPA; require multiple EIS's	DOE independent of EPA and NRC regulations, exempt from NEPA requirements for an EIS
End subsidies and federal support for nuclear waste management; require utilities to fund the repository	Waste disposal a federal responsibility; federally funded repository
End federal promotion of nuclear power	Repository needed to restore public faith in the "nuclear option"
Rely on congressional or public determinations of the need for disposal facilities	Administrative (DOE) determination of need

by past commitments to nuclear power or a specific waste disposal technology.

The nuclear establishment's commitment to placing a high-level waste in a deep geologic repository was early and essentially continuous. It outlasted proposals that waste be dropped into the Antarctic ice shelf, blasted into space, sent to a remote Pacific Island already contaminated by nuclear weapons tests, or buried in seabed sediments.[2] Federal expenditures for commercial waste management increased from $1.7 million (1972) to $317 million (1982). By 1981 DOE was investing approximately $270 million a year in commercial high-level waste management, and most of this money was committed to work on the repository solution.[3] The geological, hydrological, and other studies needed to ascertain site-specific conditions actually commenced years before the 1982 NWPA. Reports to Congress showed that by 1978 substantial on-site work had begun in Texas' Permian Basin, Utah's Paradox Basin, and the Gulf Coast salt domes. Work at other sites was progressing too: "In FY-1977, the Nevada Nuclear Waste Storage Investigations Project was formally organized by the DOE to investigate and determine whether specific underground rock masses are suitable for permanently disposing of highly radioactive wastes and to determine whether the NTS [Nevada Test Site] would qualify as a suitable repository site."[4] By 1979 attention had narrowed to the tuff of Yucca Mountain. Similarly, early studies were completed as part of the Basalt Waste Isolation Project on the Hanford, Washington, nuclear reservation. Much of the equipment needed to begin sinking mine shafts had already been purchased and was on site by the time the 1982 NWPA was passed. Thousands of pages of bibliographies were evidence of the immense number of engineering and site studies already completed by DOE contractors.[5] A huge scientific and financial investment had already been made in the repository solution, but assessments of the value of this investment varied. In essence, the question was whether this investment represented a body of credible scientific research or if it was simply a mass of unreliable data and analyses that could be used to rationalize the push for a quick solution.

At a 1976 congressional hearing an assistant director of the FBI asserted that the Communist party has "a program to try to discourage the use of nuclear energy in the United States." Pro-nuclear members of Congress dismissed scientists critical of nuclear power as anti-American cranks and "extremists."[6] Nevertheless, the attacks on AEC and DOE science gained recognition and credibility, especially when they revealed instances of suppressed studies and data on reactor accidents or the health effects of radiation. An unwillingness to release sensitive information and allow public scrutiny of its research increased suspicions that the science used to support AEC-DOE decision making was less than reputable.[7] Many scientists questioned DOE's confidence in mined repositories and viewed the agency's choice of geologic formations as premature. They criticized the agency's confidence in the ability of a repository to isolate waste from the biosphere and its reliance on simplistic models of questionable validity. The geologist who had revealed the flaws in AEC's plans for a repository at Lyons, Kansas, called on Congress to initiate a program of true research and development, not simply an engineering demonstration project. Groups such as the Natural Resources Defense Council called on Congress to create an independent scientific body to review DOE's science.[8] On the other side, as Congress approached passage of the 1982 NWPA, the secretary of energy pointed to over twenty years of research and study as a substantial investment, and saw no reason for additional site screening, research, or testing, which would slow implementation of a technological solution.[9]

Questions about the reliability of repository technology and alternative waste management schemes multiplied as DOE and Congress converged upon repository locations in Utah, Nevada, Washington, Texas, and the Gulf Coast. Members of the nuclear establishment vacillated among the various schemes and proposals. Some utilities continued to question the definition of spent fuel as waste and proposed long-term but retrievable storage that would permit owners of spent fuel to eventually recover and sell the uranium, plutonium, and other isotopes. Others thought that DOE should reprocess spent fuel and stockpile the plutonium to feed the breeder reactors of the future. Critics suspected that the Reagan administration's decision to revive reprocessing of commercial spent fuel was part of a plan to use commercial reactors to supply the plutonium for expansion of the U.S. nuclear arsenal.[10] Commercially, it made no sense to reprocess spent fuel, especially after the uranium cartel collapsed and prices went into a nose dive.[11] Some used the technical uncertainties of permanent disposal to propose a monitored, retrievable storage facility (MRS) over a repository—a position championed by Senator J. Bennett Johnston of Louisiana. Johnston's proposed MRS facility could be located virtually anywhere in the country since it did not depend upon local geology, such as Louisiana's salt domes, to isolate radioactive waste from the biosphere. Proposals were advanced to validate or discredit past analysis of repository sites. Critics pointed out that past procedures and criteria which were used to screen sites were neither publicly reviewed nor systematic, and they called for scrapping the DOE's ad hoc process. In its place, they demanded federal legislation which would require specific site selection criteria and procedures for public comment and review. They demanded a national screening program free from politics and past commitments.[12] Under this scheme, DOE could not defer its research on granite (favored in many European studies), favor sites on federal nuclear reservations, or defer consideration of sites in the eastern United States.

The data used in the screening of potential repository sites before passage of the NWPA was criticized by many, including the U.S. Geological Survey (USGS), which observed that: "it is highly unlikely that sufficient earth-science information will be available by June 1, 1982 or June 1, 1983, for three candidate sites to be selected objectively. With inadequate information on which to base the decision, DOE will be forced to "guess" which localities may offer the best chance of success. It also appears to increase the risk that the final site selection may prove to be unacceptable."[13] Legislation mandating the quick identification of potentially acceptable sites, without requiring national screening on the basis of key geological and hydrological criteria, would "lock you into sites you already know the most about." More important, in the opinion of the USGS, the Department of Energy had used simplistic criteria to screen sites on the basis of geological formations (for example, salt, basalt) rather than considering key geohydrologic properties that would inhibit the transport of radionuclides, via groundwater, out of a repository. In 1982 a spokesman for the USGS said that existing hydrologic data was "inadequate" for DOE to "judge the suitability of the area [in Utah] for a high-level waste repository."[14] Moreover, the same spokesman warned that the prototypical and unpredictable nature of this experimental repository technology would produce unexpected delays and technical surprises that should be allowed for when setting a schedule for the waste management program.

But most of the proposed legislation assumed that the science and methods already used to identify potential sites could be grandfathered into a congressionally mandated site selection process. Once again the fear of delay was used to justify an abbreviated process of site screening and selection. The Department of Energy claimed that consideration of new sites, apart from those already identified, would produce at least a four-year delay in the repository program.[15] The Reagan administration and key supporters of the nuclear establishment in Congress wanted legislation that would put an end to the "paralysis of analysis" and end disputes over the best technology or location. Just as congressional action which mandated scrubbers decided how the nation would reduce sulfur dioxide emissions, so an act of Congress which mandated a repository at an already studied location could end the discussion of alternative solutions. If a repository was adopted as the final and permanent solution, future political discussion about the merits of other technologies—such as the surface storage facilities being considered in Europe—could be foreclosed. The certainty of science could replace the uncertainty and doubts raised by critics of the nuclear establishment who called for new research and development programs. With the political and capital commitments made, it would be a solution difficult to abandon.

Proposals which supported the adoption of existing technologies and already studied sites favored the scientific and decision-making authority of the nuclear establishment. Debate thus could be closed on the appropriateness of other technologies or site selection criteria. The primacy of existing institutional expertise could be established, there being no need to consider an alternative set of technical or planning assumptions. The continuity of past research programs and contracts with the companies conducting that research could be assured. Similarly, proponents and potential beneficiaries of the repository solution (for example, companies with experience drilling large diameter mine shafts) could continue their association with other members of the nuclear establishment. In short, endorsing the technical expertise of the nuclear establishment was viewed as a key to accelerating the construction of a repository and resolving a major problem confronting the future of nuclear power.

Addressing Questions of Institutional Legitimacy

Love Canal, Times Beach, the leaking Hanford tank farm, Lyons, Kansas, and West Valley, New York, were more than examples of bad technical judgments. For many people, they were evidence of institutional failures and reasons to distrust the state's and industry's commitment to protecting public health and safety. Books and newspapers regularly reported on leaking waste storage facilities which the public had been assured were safe and trouble-free when constructed. The disastrous record of both chemical and radioactive waste management pointed to the risks of depending upon institutional surveillance and maintenance of temporary waste storage facilities. Isolating high-level waste from the biosphere for thousands of years presented a particularly acute problem, the life span of which far exceeded the stability of known institutions. Hence, the attractiveness of a solution—the permanently sealed repository—would be free from institutional maintenance and monitoring. Reinforced by a history of industry and state mismanagement and distrust of institutional solutions, the Environmental Policy Institute, the Sierra Club, and other groups advocated a permanent, irreversible technological solution.[16] For critics of the nuclear establishment, a lasting solution reduced the possibility of a radiological disaster due to mismanagement or the retrieval of spent fuel for weapons production or breeder reactors. The nuclear establishment also looked to a permanent resolution but obviously for different reasons. A lasting settlement would eliminate waste disposal as an argument against nuclear power and free the industry from the vagaries of government policies and priorities. Even if the permanent solution was widely accepted as desirable, its implementation could still be disrupted by conflicts over expanding state-local authority, demands for greater public participation and oversight, and calls for an entirely new waste management agency.

Opposition to continued DOE control of waste management came down to a conflict of interest question. Could the agency responsible for promoting nuclear energy also be responsible for protecting public health and safety? This same conflict—which had led to disbanding of the AEC—suggested that a new institution solely dedicated to managing high-level waste was a prerequisite to a legitimate repository program.

The Interagency Review Group initiated by President Carter had attempted to develop a consensus among federal agencies on the institutional requirements of a repository program. Unfortunately, the group was chaired by the secretary of energy, and its conclusions were criticized as being watered down to satisfy the DOE. Nevertheless, the Interagency Review Group issued a report which called for a federal "super-board" to oversee and resolve conflicts over high-level nuclear waste management. Such a board could expand the base of expertise and range of interests brought into the waste management debate beyond those offered by the nuclear establishment. Other proposals offered included: creating a federal waste management corporation; a quasi-independent oversight commission composed of federal and state officials; a waste management office within EPA; and other sorts of institutional arrangements.[17] The variety of proposals reflected a basic conflict over the role of the state. Opponents of the nuclear establishment justified state involvement in terms of its function of protecting public health and safety. However, for the nuclear establishment, the major goal was to protect investments in nuclear power plants and to keep nuclear industries viable. From the latter point of view, waste management fit within DOE's mission and structure.

Apart from environmental and civil groups, the Office of Technology Assessment (OTA) was one of the most forceful critics of past waste management programs and a proponent of institutional change. In the words of OTA, "The deepest and most consuming doubts concern not so much the technology for disposal . . . but rather the institutional capacity of the federal government to carry out, over a period of twenty years or more, the difficult and sustained effort required to both build and operate a disposal system capable of handling large amounts of high level radioactive waste."[18] In OTA's opinion, the Department of Energy was unsuitable for the job: "An independent institution with independent funding does seem the surest—perhaps the only—way to guarantee that a comprehensive waste management program will be carried out from start to finish and on schedule."[19] The OTA saw the technical and institutional elements of nuclear waste policy as mutually supportive and inseparable. It criticized proposals for speeding the selection and licensing of repository sites that left insufficient time to resolve important institutional issues. In OTA's opinion, forcing the quick implementation of siting decisions would only produce substantial and sustained opposition to the DOE program.[20] In 1982, only months before passage of the 1982 NWPA, the Office of Technology Assessment still believed that the legislation inadequately defined institutional arrangements critical for the success of a waste management program.[21]

At this same time other groups were calling for the whole-scale reconstruction of waste management programs and institutions. According to Representative Al Swift of Washington, DOE's credibility with the states was at a historic low point. Calling a site selection process by DOE a "dumb show," he called for an independent commission to assure the scientific credibility of the site selection process and "to bring in people who you can less likely accuse of having a bias or predetermined judgments from having massaged this material for so many years."[22] Representative Ron Wyden of Oregon proposed:

An independent review body that would, in effect, monitor the DOE with respect to site selection and site characterization.

What we would be interested in is a jury of sorts which would set out criteria which DOE would use in characterizing sites and also in terms of the approval or disapproval process. In effect they [the jury] would be submitting a set of final sites for Presidential selection.[23]

In this proposal, DOE would be reduced to serving the discretion of "the jury." Representative Stanley Lundine (New York) also favored this approach in a proposal which called for the establishment of a permanent, independent mediation-arbitration board made of representatives of Congress, the states, industry, and environmental groups. The board would hear and rule on complaints and charges against the DOE. A few years before, after the Three Mile Island accident prompted widespread criticism of the NRC and other members of the nuclear establishment, Representative Morris Udall proposed a comprehensive revision of the Atomic Energy Act. One part of that proposed legislation called for the establishment of a nuclear safety board as an independent agency (similar to the National Transportation Safety Board) which would investigate accidents and the management of nuclear facilities. Its purpose was to assure the public that an agency independent of the nuclear establishment would be available to offer an impartial assessment of nuclear safety and environmental violations and accidents.[24]

From DOE's point of view, new institutional arrangements under a nuclear waste policy act were unnecessary because: "Most of the authority and much of the institutional structure for implementing our waste management strategy is already in place. DOE's responsibilities for waste management are specified in existing legislation. . . . In view of the framework that is already in place, I would caution against legislative action . . . to prescribe additional layering of responsibilities.[25] The only additional mechanisms DOE believed it needed were ones for ending state opposition and negotiating compensation with local governments.

Up to this time it was never clear what waste management decisions would be made when, or on what basis. Little advance warning was given about the release of key documents or decisions. In many cases outsiders were trapped in a circular logic: they did not know about a decision unless they requested specific information from DOE, but the department was under no obligation to publicly announce its program decisions. This arrangement left DOE with much latitude to adjust program schedules, priorities, and to shield its operations from public scrutiny.[26] The congressional Office of Technology Assessment, and others dissatisfied with this lack of institutionalized procedure, recommended that DOE be required to submit a "mission plan" and project schedule to Congress for approval.[27] A mission plan would force DOE to declare its program goals, identify key program components, and establish a schedule of important decisions. It would require DOE to explain its repository program and justify program changes in updates submitted to Congress. In response, DOE maintained that through the budget authorization and appropriations processes, Congress already had adequate oversight of DOE's activities. A mission plan was just another unnecessary requirement.

Concerned about proposals for dismantling DOE, industry supported the reassertion of DOE's control and called for legislation that would confirm its status as the lead management agency. Nuclear industries and utilities opposed proposals to establish an arbitration board, an interagency review group, a state veto, or a state planning council.[28] The State Planning Council was originally established in 1980 by President Carter to strengthen intergovernmental relationships and give states "better access to information and an expanded opportunity to guide waste management planning."[29] However, plans for a state planning council disappeared from legislation supported by the Reagan administration.

Some arguments against institutional change centered on political instability as a major reason for the lack of a coherent approach to the problem. At least three major reorganizations of federal energy agencies occurred between 1972 and 1982 along with major changes in waste disposal policies ordered by presidents Carter and Reagan. Critics of the nuclear establishment who proposed yet another reorganization raised the specter of additional delay, bureaucratic instability, reorganization costs, and conflict over the shape of that new institution. Creation of another federal agency also contradicted the Reagan administration's rhetoric about reducing the federal bureaucracy. More important to the nuclear establishment, a wholly new and independent institution looked like another EPA on the horizon—yet another threat to its discretion, authority, and business practices. To its way of thinking, the availability and cost-effectiveness of relying on resident expertise and existing institutions were commanding justifications for retaining DOE management. Continuing with DOE offered the easiest path to a near-term demonstration that something could be done about the high-level radioactive waste problem.[30]

Realigning Federal, State, and Local Power

Seeking to limit state and local interference, members of the nuclear establishment regularly called for continued and expanded federal preemption of all decisions related to nuclear power and waste management. Utilities found that the waste disposal issue was often raised in challenges to new reactor licensing. The NRC was forced to initiate the Waste Confidence Proceedings to assess the feasibility and availability of a facility to receive spent fuel from future commercial reactors slated for operating licenses. In the first year of the Reagan administration, DOE aggressively sought to limit outside interference in its waste management programs by calling on Congress to

> modify the proposed bill [S.1662] to delay presidential and congressional involvement. . . . Allow the Secretary [of Energy] to

be "arbiter" of any objections to the recommendation of candidate sites for site characterization. . . . Explicitly disallow any further encumbrance upon the activities of the program by a governmental body, beyond the requirements for DOE to consult with the States.[31]

In essence, DOE was asking for unlimited discretion over waste management programs and the authority to rule on the validity of state and local complaints. The department had long supported legal challenges of state and local regulations by nuclear industries. Responding to a request from President Reagan, the secretary of energy consulted with the nuclear industry to determine which regulatory barriers were hindering their expansion and operations.

As another way of circumventing state-local restrictions, and avoiding future state-federal confrontations, members of the nuclear establishment endorsed the construction of a pilot waste disposal facility designed to nullify restrictions on power plant development due to the lack of a waste disposal facility. Proposed federal legislation was explicitly discussed in terms of the question—will this facility be enough to satisfy the California requirements?[32] If defined as an experimental DOE-owned facility, a pilot repository would automatically be exempt from NRC licensing, the National Environmental Policy Act, and state oversight. Given the restricted opportunity for public opposition and judicial and state review, DOE estimated the facility could be ready to receive spent fuel from utilities within seven years. Proponents of the pilot facility in the House Science and Technology Committee asserted that the technology was available, but "there are many institutional and political problems that stand in the way of demonstrating the availability of this technology to the American people. *This bill is designed to minimize political barriers* and set in place a program to demonstrate that the technology exists for safe and permanent disposal" (emphasis added).[33] Defining the facility as a demonstration project would also limit other (more critical) committees' jurisdiction over the project. Utilities welcomed the possibility of a pilot facility: "Closure of the back end of the fuel cycle is very important in assuring the public that radioactive wastes are manageable. Until the public has such an assurance there will be concern about expansion, or even continued operation, of nuclear facilities."[34] Some proposals went even further and gave DOE the discretion to convert the demonstration facility into a full-scale repository without completing an environmental impact statement or involving the host state. If followed through, these proposals would have given DOE complete authority over all aspects of repository siting, design, and operation. Procedural demands imposed by NEPA would be circumvented and DOE would be permitted to proceed uninhibited by state, local, and other groups.

Organizations such as the National Governors Association sought to limit DOE's discretion and expand state power in two ways: (1) by requiring some form of consultation and concurrence in which DOE would be required to reach a consensus with the states—rather than simply informing them of its decisions; and (2) by giving a state the power to veto a siting decision. The state veto was the most controversial proposal. As originally proposed by the association and endorsed by groups such as the Union of Concerned Scientists, the state veto would have been absolute. No state would have been forced to accept a repository. However, within the NGA there was only limited support for an absolute veto. States which wanted to get rid of waste foresaw a deadlocked repository program and reactor sites as de facto repositories. States with little chance of being considered for a repository, but who had a stock of high-level waste awaiting disposal, opposed a strong state veto. Other proposals suggested a veto, actually a notice of disapproval, which would be either sustained or overridden by one or both houses of Congress. The nuclear establishment opposed any form of a state veto. But when passage of the NWPA depended upon offering the states some form of a veto, debate turned to the issue of congressional approval.

The nuclear establishment proposed requiring both houses of Congress to pass a resolution sustaining the state veto and requiring that the resolution be signed by the president. This last requirement would require the president to reject the site recommendation of his own secretary of energy. Under this form of veto, the burden of proof would rest with the state (not DOE) to defend its position before Congress. States sought the most leverage by requiring an unconditional veto, or at least a veto that would stand *unless* overridden by a majority vote in both houses of Congress.[35] However, some groups outside the nuclear establishment, such as the Natural Resources Defense Council, opposed any veto on the principle that it would "politicize" a site selection process that, in their opinion, should be strictly objective and scientific.

Proposals for consultation, concurrence, and cooperation between federal and state governments were similarly varied. Some states believed that when a state-federal conflict occurred the program would be halted until the problem was negotiated to agreement. Again, many saw such an arrangement as giving the host state too much power. Alternate proposals called for a formal agreement which would define a process for resolving DOE-state conflicts, but would not give a state the power to halt the repository program.

Public Involvement

Demands that the repository program incorporate multiple opportunities for public involvement were countered by proposals to give the secretary of energy final authority to define public involvement. One potential repository state, Mississippi, said that Congress could ill afford to leave resolution of issues "to chance through the presently unpredictable procedure of agency regulations devoid of policy direction."[36] Formal channels for public involvement were demanded. Environmental groups advocated multiple opportunities for public participation in site selection, characterization, construction, and operation of the repository. In their view, the way to facilitate this was to require diligent application of the NEPA standards. States with potential repository sites feared that site characterization studies (which included the construction of mine shafts) would impose huge environmental and socioeconomic burdens on local jurisdictions. They demanded an environmental impact statement (EIS) and adherence to NEPA as crucial for determining the extent of socioeconomic impacts and a fair level of compensation. Even more important, an impact statement would provide an early definition of significant issues and an indication of DOE's willingness to respond to state-local interests. DOE, however, preferred to delay application of NEPA requirements until after millions of dollars had been invested in site characterization. At that point, an EIS would only apply to the final site selection decision— too late to make a difference.

Despite vocal criticism of its public involvement program, DOE asserted that informal cooperation with the states was sufficient and that no formal requirements for coordination and consultation (let alone concurrence in program decisions) were necessary. When questioned about its reaction should a state refuse to cooperate, DOE cited its authority under the Atomic Energy Act to override the opposition and take any action it felt was necessary to implement its policies.

For the local level, the National Association of Counties proposed that new means of participation should include local jurisdictions, especially since counties in likely transportation corridors to a repository were already passing ordinances on nuclear waste transport. Counties near potential repository sites also had a major stake in site selection, impact mitigation, and compensation. In testimony before Congress, the association expressed its concern that information would not filter down from the state to the local level, and that local people would not be included in interactions with federal agencies.[37] Local citizen groups were also concerned that their participation would be forgotten in the battle over the role of the host state. When it was proposed that states would receive funding to allow for monitoring the DOE program, public interest groups maintained that they should also receive funding. Just as utilities opposed funding interest group involvement in the reactor licensing process, they opposed such funding as part of the repository program:

> The utility industry and others have consistently opposed this type of funding on the principle that it would not contribute to effective and efficient proceedings. . . . Intervenor funding would only result in further delays by granting financial assistance to intervenor groups many of whose primary purpose is to delay and prevent nuclear power plant licensing, rather than improve their safety.[38]

Opposition to broad forms of public involvement occurred despite the fact that citizen and environmental groups had a long history of participation in nuclear waste issues. The Environmental Policy Institute and the Natural Resources Defense Council, for example, had testified at nearly every major congressional hearing on nuclear waste policy. At various times they lent their expertise to members of Congress, local governments, and had joined with states in challenging the Department of Energy.

Whatever its institutional form, a repository program would have to be administered. Heretofore energy agencies (AEC, ERDA, DOE) had always supervised nuclear waste programs. However, their record of confrontation with state, local, and civil groups; their loss of scientific credibility; their attempts to limit state and public involvement; and their obvious support for the interests of nuclear industries and utilities all suggested that there would be serious consequences to foregoing institutional reform. Restoring public faith in the institution responsible for waste management was crucial to the program's success. Reassertion of DOE's administration would have to confront its legacy of distrust and the difficulty of restoring public faith in its management of a nuclear waste program.

Addressing the Financial Crisis

Advocates of reconstruction did not assume that a nuclear waste program should be tailored to address the economic problems of the nuclear industry. In their view, the financial crisis of the nuclear industry was simply evidence of the failure of nuclear power to pass the market test. Secondly, this crisis was the fabrication of an industry suffering from mismanagement and in search of yet another federal bailout. For years DOE and the utilities had used the scenario of filled reactor storage pools leading to reactor shutdowns and power shortages to defend the need for development of a repository. But when would this crisis materialize? The date had been pushed back so many times that it appeared neither imminent nor credible. Even defenders of DOE and nuclear power, such as Senator James McClure of Idaho, questioned the credibility of DOE projections. In March 1980, DOE said 755 tons of storage capacity would be needed by 1985. A year later, it said no additional storage would be needed by that date. McClure, who had used DOE's 1980 projections to defend his support for the rapid development of nuclear waste facilities, was "surprised" by the new estimates.[39] In 1978 the Atomic Industrial Forum grossly overestimated that the loss of nuclear-generating capacity due to insufficient spent fuel storage would be 660 megawatts by 1981, fourteen hundred megawatts by 1983, and an incredible ninety-five hundred megawatts by 1988. Governor Riley of South Carolina observed: "It is argued that federal AFRs [temporary, away-from-reactor storage facilities] are needed in order to prevent reactors from shutting down operation in the late 1980s. But every year for the past three years the official estimates of how much capacity will be needed in federal AFRs have dropped dramatically. And every year for the past three years the official estimate of when this capacity would be needed have stretched further and further into the future."[40] By 1982 no nuclear power plants had been shut down due to lack of spent fuel storage, and in the next decade none appeared

likely to be shut down for that reason. Critics who bally-
hooed the crisis pointed to TVA as an example of how utilities could take care of their storage problems on their own.
(The TVA did not support plans for a federal temporary storage facility.) In short, there was no crisis justifying accelera-
tion of the site selection program. And there was no need
to use federal legislation to limit state, EPA, NRC, or judicial oversight and public involvement to speed development
of a repository. Legislation that went after the nuclear waste
issue because it threatened America's need for nuclear-
generated electricity sought out the wrong issue.[41]

Within the nuclear establishment, nuclear industries
blamed unstable policies for creating a near-term crisis in
spent fuel storage and a long-term financial crisis. West Valley, New York, was commonly cited as an example of how
an unstable political environment had destroyed the spent
fuel reprocessing industry. According to industry, the plant
was shut down, not because of technical problems, but because unpredictable changes in NRC regulations made its
continued operation uneconomical. In another sector of the
nuclear economy, lack of a stable policy was viewed as responsible for a failing U.S. uranium mining industry. In the
words of a member of Wyoming's congressional delegation,
the state's uranium mines were being closed because "there
is no market for the product. There is no market for the
product because investors have no confidence in the future
of nuclear power and they have no confidence because there
has been no clear-cut federal policy with respect to nuclear
power."[42] One major omission was the lack of a clear-cut
policy on spent fuel reprocessing, disposal, or storage. In the
rhetoric of the nuclear establishment, implementation of
a spent fuel policy was the fastest way to change public
perceptions that waste disposal was a major drawback to
the expanded use of nuclear power. Quickly putting a solution to work (such as an unlicensed demonstration facility)
would be a major step in regaining public acceptance of the
nuclear option and revitalizing the nuclear industry.

In the nuclear industry's opinion, the failure to adopt

a firm waste management policy created a near-term spent
fuel crisis by spreading distrust about the feasibility and cost
of disposal. Uncertain federal policies made it impossible
for private contractors to estimate the financial returns on
construction of a private facility for spent fuel storage. Industry argued that antinuclear groups created licensing uncertainties which limited the utilities' ability to expand
reactor-site storage. Congressional critics of the nuclear industry, such as Representative Edward Markey of Massachusetts, attacked the utilities' unwillingness to invest in developing either an interim or final solution to spent fuel
disposal. But, from industry's point of view, nuclear research
and development had always been funded by the state. They
maintained that, because basic research and development
on high-level nuclear waste disposal—like all research and
development—was for the "benefit of all citizens," it should
be paid for by the taxpayers.[43]

The nuclear industry set out two requirements for a
stable nuclear waste policy: a firm timetable for federal
waste acceptance and a predictable payment scheme that
would finalize the cost of waste disposal. Proposals for a
tax on nuclear-generated electricity were acceptable to environmental and citizen groups as an equitable means of funding a waste disposal program and avoiding another industry
bailout. Those who benefitted from nuclear power would
pay the costs. However the utilities, already faced with the
denial of some costs in their rate base, wanted to make sure
that the cost of a repository could be passed on to ratepayers. In the words of one spokesman: "I believe it will
be easier to pass such charges through to our customers if
there is an identified schedule for acceptance by the federal
government of the waste, upon payment of the defined
charge"[44] With DOE offering to enter into formal waste acceptance contracts with the utilities, the proposed tax on
nuclear-generated electricity was defined as a legitimate
business expense that could be passed through to customers.
Furthermore, a federal tax assured an equalized distribution
of financial liability within a utility's service network. If

authority to make adjustments in the tax on nuclear electricity remained at the national level, states were seriously limited in their ability to challenge the tax. Potentially controversial expenses such as financial compensation to the host state—ratepayers in New Jersey might not like watching millions of dollars flow to the state of Nevada—could be protected by the federal levy. While utilities would have preferred a one-time charge for waste disposal, they were willing to support a federal tax dedicated to a nuclear waste fund if it was tied to a "firm contractual commitment of the federal government to take title to the radioactive waste or spent fuel." The proposed Nuclear Waste Fund could free the waste management program from another source of uncertainty—federal budget cuts.[45] Utilities generally supported funding financial assistance to the states as a way to "remove one significant obstacle for locating these facilities within a state."[46]

Industry still proposed to pass on as many costs as possible. At one point it was proposed that the host state bear twenty-five percent of the cost of compensating local jurisdictions for repository-related impacts because, after all, the state's economy would reap huge benefits from the influx of federal money. However, the Office of Technology Assessment criticized DOE for assuming that the repository would be a financial windfall for the host state and for neglecting to analyze the full cost of socioeconomic impacts on local communities.[47] Legislation proposed by the nuclear establishment also avoided consideration of increased costs for improved emergency response, medical facilities, monitoring waste shipments, and vehicle inspections along transportation corridors to the repository.

Proposals for exempting waste storage facilities from NEPA and state and public review were tied to concerns about the cost of conflict and delays. The financial impact of delay was a common concern of the utilities having seen construction times for nuclear power plants extend to ten years or more. Opponents of the nuclear establishment believed that policy should not be premised on addressing the financial problems of nuclear industries. However, utilities continued to argue that federal legislation was necessary to eliminate a major threat to the continued operation and profitability of nuclear power plants throughout the country. By removing this obstacle the state would fulfill its function of protecting capital investment.

Proposals for Reassertion or Reform

An environment of disorder and change presented an opportunity to reexamine U.S. commitment to nuclear energy in light of the difficulty in accomplishing an effective program of nuclear waste management. Support for reform could be found at all levels of state and civil society. In the opinion of many, the credibility, legitimacy, and financial crises were evidence of a need to define new roles, institutions, procedures, and relationships. Options for reconstructing institutional relations, and creating new institutions responsible for addressing the problems of high-level waste disposal, were offered by a diverse set of political groups and analysts; nuclear waste legislation gave them an opportunity to accomplish reform and reconstruction.

Countering suggestions for reform or reconstruction were proposals which relied on the authority of the national state to restore the viability of the nuclear establishment— its industries, government agencies, and science. Reasserting the nuclear establishment meant restricting the basis for political challenges to its authority, decisions, expertise, and control over nuclear affairs. In short, proposals were advanced with the aim of denying its political opponents a basis from which to contest the legitimacy, credibility, and fiscal responsibility of the nuclear establishment.

CHAPTER 5

▼

THE FAILURE OF THE 1982 NUCLEAR WASTE POLICY ACT

THE 1982 NUCLEAR WASTE POLICY ACT (NWPA) is a difficult and ambiguous document. While it outlined a new site selection process (table 4) and increased opportunities for state involvement in the repository program, it did little to resolve concerns about the credibility, legitimacy, or financial condition of the nuclear establishment. Long-standing conflicts between the nuclear establishment and its opponents remained. The act, analyzed in light of historical precedent, is hardly revolutionary. In spite of some innovative approaches found within it, the act reflected a general unwillingness to challenge the interests and authority of the nuclear establishment.

On the balance, the 1982 NWPA offered few concessions to groups and individuals challenging the nuclear establishment. It produced only minimal changes in state policies for subsidizing and promoting nuclear energy. The NWPA did not redirect the goals and functions of state agencies, nor did it establish the superiority of the state's role to protect public health over that of serving the interests of the nuclear industry and protecting its capital investments. Ideally, legislation of the nation's first high-level nuclear waste policy was an opportunity to create a more open decision-making process within which innovative and more credible approaches to nuclear waste management could be explored. In practice, it left the goals and course of past waste manage-

TABLE 4

STATUTORY DEADLINES MANDATED BY THE
1982 NUCLEAR WASTE POLICY ACT

Sight Selection Guidelines and Program Planning

7 April 1983	DOE identifies states with potentially acceptable sites (nine sites identified)
7 July 1983	DOE issues proposed site selection guidelines
1 January 1984	NRC issues technical guidelines
8 January 1984	EPA issues standards for off-site releases
6 April 1984	DOE issues draft Mission Plan
6 July 1984	Final Mission Plan sent to Congress
7 January 1985	President must decide if repository may also be used for military waste

Site Selection: First Repository

1 January 1985	Secretary of Energy recommends three sites for the first repository for characterization. Environmental assessments prepared prior to recommendation
1 March 1985	President approves three sites for characterization—first repository
31 March 1987	Presidential recommendation on the first repository site sent to Congress
31 May 31 1987	Deadline for first host state to submit its notice of disapproval (or sixty days after the presidential recommendation)
1 January 1989	NRC decision on DOE's application to construct the first repository (or within three years after its submittal)

Site Selection: Second Repository

1 July 1989	Secretary of Energy recommends three sites for the second repository for characterization
1 September 1989	President approves three sites for characterization—second repository
31 March 1990	President submits a recommendation for the site of the second repository to Congress
31 May 1990	Deadline for second host State to submit its notice of disapproval (or sixty days after presidential recommendation)
1 January 1992	NRC decision on DOE's application to construct the second repository (or within three years after its submittal)

ment programs unaltered and the authority of the nuclear establishment largely intact.

To be fair, one must recognize that the act introduced new and unprecedented procedures and powers into the operation of nuclear waste programs. It provided a limited state veto over the final site selection decision; recognized that the rights of native American nations were equal to those of the states; and required consumers of nuclear electricity to pay for the repository. It required adoption of explicit site selection and environmental protection criteria, that a final site be selected on the basis of characterization studies, and that a plan for the repository program (the Mission Plan) be submitted for congressional review.

These concessions amounted to much less than they might have had an alternative institutional structure been set up. With the institutional structure of the nuclear establishment largely intact, the DOE was free to redefine the administrative manifestation and significance of these innovations. Without major institutional revisions, the NWPA could only be implemented as a conservative and backward-looking piece of legislation with the effect of prolonging the conflicts associated with old institutions and abbreviating attempts to define new ones more responsive to contemporary political and social demands.

A hodgepodge of contradictory and incompatible stipulations, the act only materialized in the final, hectic hours of the Ninety-seventh Congress. The Reagan administration and the nuclear industry pushed hard to finalize the NWPA in 1982. In the lame duck session, nuclear waste had to compete for attention with thirteen appropriations bills and controversial legislation on social security, the Clean Air Act, immigration reform, hazardous waste, and tuition tax credits. Even setting aside time constraints, passage of any nuclear waste legislation involved the complex task of satisfying seven committees in the House of Representatives alone. A menagerie of committees and subcommittees—on the budget, judiciary, commerce, science, energy, interior, and armed forces—claimed jurisdiction over nuclear issues.[1]

The armed services committees demanded that military waste be exempt from the act's provisions. Various interior subcommittees opposed provisions for federal funding of away-from-reactor, interim, or monitored retrievable storage, seeing them as bailouts for industry and distractions from the goal of permanent waste disposal. Committee chairs, such as Representative Richard Ottinger (Ohio) opposed exclusion of repository siting from National Environmental Policy Act requirements, and attacked the DOE and the nuclear industry for insisting on unrealistic legal requirements.

At the end of September 1982, three months before passage of the NWPA, representatives were still arguing over basic issues such as permanent versus temporary disposal, and whether high-level military waste would be stored with civilian waste. The legislation that finally emerged reflected the inadequacies of a last-minute compromise worked out between the House Committee on Interior and Insular Affairs, chaired by Morris Udall, and Senator McClure's Energy Committee. On 20 December 1982, compromise legislation was passed with little debate—twenty amendments were disposed of in as many minutes. A joint committee report, which may have clarified the goals and intent of those crafting the compromise bill, was never issued.

In the author's opinion, the 1982 act failed to accomplish the following four changes in the structure of the nuclear establishment, which were needed to address the credibility, legitimacy, and financial crises.

Failure to Redefine the Goals of State Institutions

Challengers of the nuclear establishment wanted national recognition that the primary goal of nuclear waste management was to protect environmental health and public safety. Few would disagree that health and environment were important, but a more compromising stance was incorporated into the NWPA. Rather than outright endorsement, the act hedged a commitment to the goal of public health and safety. Protection of health, environment, and safety are not mentioned in the act's preface; rather, the purpose of the repository program was "to establish a schedule for the siting, construction, and operation of repositories that will provide a *reasonable* assurance that the public and the environment will be *adequately* protected from the hazards posed by high-level radioactive waste" (emphasis added).[2] Congress went so far as to recognize that "radioactive waste creates *potential risks* . . . [that] a national problem has been created by the accumulation of (A) spent fuel from nuclear reactors; and (B) radioactive waste from (i) reprocessing of spent nuclear fuel. . . . *Federal efforts* during the past 30 years to devise a permanent solution to the problems of civilian radioactive waste *have not been adequate*" (emphasis added).[3] The responsibility of the federal government to protect public health, safety, and the environment was recognized only in contrast to industry's responsibility to pay for the costs of waste disposal. Was the goal to dispose of nuclear waste as quickly, and as safely, as possible? Or was it to dispose of nuclear waste as safely as possible? Emphasis on meeting a waste acceptance schedule raised additional questions about the trade-offs between economics and health when defining "reasonable" assurance, "adequate" protection, and real versus "potential" risks.

The less-than-paramount priority attached to protecting public and environmental health is also evident in the exemption of site characterization decisions from NEPA requirements. Eventually it was decided that the choice of locations for site characterization would be based on an "environmental assessment" (EA) rather than a full environmental impact statement (EIS), which would be typical for such a project. Site characterization was a multi-billion-dollar construction project that involved building mine shafts, thousands of feet of drifts, haul roads, tailings piles, and drilling thousands of exploratory boreholes. In short, it was a major underground mining operation which, if sponsored by a private company, would have been subject to an extensive EIS. Given past attempts by Congress to exclude repository decisions from NEPA, the scope and standard of evidence for the environmental assessment was ambiguous and subsequently was contested in lawsuits filed by several states and environmental groups. The compromised nature of the EA was evident in the requirement that it need only be based, at the discretion of the secretary of energy, on "available geophysical, geologic, geochemical and hydrologic, and other information."[4] At potential repository sites, such as Davis and Lavender canyons in Utah's Paradox Basin, available information was incomplete, out of date, and unreliable according to the state's scientists. Even when challenged, the NWPA limited judicial review of the environmental assessment to criteria spelled out in the act and site selection criteria defined by DOE. A complete review of environmental impacts under NEPA standards would not occur until billions of dollars had already been spent for characterizing three sites and after DOE had selected one of those sites for the repository.

The decision to turn the program over to the DOE—an agency many viewed as uncommitted to enforcing health and environmental standards at its own facilities—was an endorsement that such concerns would have to share the agenda with the economic interests of the nuclear establishment.[5] The act focused on streamlining the site selection process, limiting state and local interference, and speeding up the construction of a repository. Key members of Congress emphasized that they did not want program delays to plague the new repository program. They asked the NRC and DOE to cooperatively determine what information would be absolutely necessary for repository licensing so that the licensing process could proceed as smoothly as possible.[6]

The Nuclear Regulatory Commission, the federal agency responsible for protecting public health and safety, emphasized that it structured its licensing procedure to "expedite the formal licensing proceeding by providing for an informal pre-licensing process. Under this process, the interested public and NRC staff could call attention to important issues and settle on measures needed to resolve them *without encumbering DOE's flexibility"* (emphasis added).[7] Under the NWPA, the EPA was neither mandated nor forced to assume a more aggressive role in overseeing DOE's handling of nuclear waste. Moreover, the Reagan administration EPA delayed issuing environmental standards for a repository, notwithstanding that by 1982 it had been working on them for five years. In congressional testimony, a meek and impotent EPA voiced its confidence in the quality of DOE and NRC programs even as the NRC threatened to override EPA and establish its own standards for permissible releases of radioactivity from an operating repository.[8] The EPA was similarly quiet on congressional proposals to exempt much of the high-level waste management program from NEPA requirements. The NWPA failed to put in place the mechanisms for aggressively applying environmental standards to the repository. Congress ignored the credibility issue raised by opponents of the nuclear establishment who charged that giving responsibility for waste management to an agency whose mission was to promote nuclear energy would compromise the protection of public health and safety. Congress also obviously ignored expressions of the public's distrust that the EPA, NRC, and DOE would rigorously enforce such standards.

Missing from the NWPA was another issue often raised in the mass media—the advisability of continuing to use nuclear energy, given the nuclear waste problem. In 1980, Senator Gary Hart of Colorado wrote:

> If a repository is not opened by January 1, 2000, the NRC must suspend issuance of new operating licenses and construction permits until a repository is opened.
>
> The committee believes this provision underscores the substantial national interest in determining whether the current projection for opening a repository can be met. . . . Moreover, it accords with the notion that the Nation cannot continue to expand its reliance on nuclear power if, within the next 20 years, the Federal Government has not implemented a safe, reliable method for disposing of the resulting wastes.[9]

An earlier omnibus nuclear bill, drafted by Representative Morris Udall, attempted a comprehensive reassessment of the nuclear fuel cycle, linking the waste management issue to the construction of additional nuclear power plants. Reactor decommissioning was a waste disposal problem often discussed in *Science*, NRC documents, and other publications.[10] But the relationship between waste production and disposal was dismissed with the argument that the disposal problem existed even if no new waste was generated. Thus, the NWPA advanced no assessment of whether the nation should continue to produce such a controversial by-product. Similarly, the act stated no position on whether the state should continue to approve and license additional nuclear facilities that produced the waste. Bypassing questions of state and institutional purpose, the NWPA suggested no fundamental reorientation toward a greater emphasis on environmental protection and public health. Waste disposal was to be dealt with as a problem isolated from other aspects of the nuclear fuel cycle.

The Distribution of Authority Remains Unchanged

The NWPA made one addition to the institutional landscape—the DOE's Office of Civilian Radioactive Waste Management (OCRWM)—but made no substantial change in the distribution of power. OCRWM was responsible for carrying out the tasks required of the secretary of energy under the NWPA and was subject to the secretary's supervision.[11] Before it could begin construction or operation of the repository, OCRWM was required to submit license applications to the Nuclear Regulatory Commission. However, the site selection process continued to operate outside the formal purview of the NRC and other federal agencies. The Department of Energy was required to consult with other federal agencies in promulgating site selection guidelines but was under no obligation to incorporate their suggestions. The Nuclear Regulatory Commission reviewed these guidelines but otherwise would not take an active role in reviewing sites. Having opposed congressional attempts to extend its authority to include approving sites for characterization, the NRC was content to concentrate its oversight activities on DOE's application for authorization to construct the repository which at the earliest, will be submitted sometime in the 1990s.

Within one year after passage of the 1982 NWPA, the EPA was required to: "promulgate generally applicable standards for protection of the general environment from offsite releases from radioactive material in repositories."[12] This decision was specifically exempted from review under the NEPA. However, draft standards were not issued until 1986 and then only after lawsuits were filed against the agency by the Natural Resources Defense Council. Thus, the EPA standards were largely irrelevant to DOE's site selection process. There was an Office of Environment within DOE, but a study by the Government Accounting Office concluded that it had little involvement in agency decision making.[13] In general, other agencies continued to have little or no authority over the repository program after passage of the NWPA.

Unlike earlier proposals advanced by the National Governors Association that called for a process of concurrence with site states, the NWPA only recognized the need for consultation.[14] Concurrence implied that states would have multiple opportunities to disapprove of DOE's actions and see that conflicts were resolved before the program continued. Consultation and cooperation emphasized public information programs. The NWPA rejected concurrence or any form of binding arbitration in favor of a process of consultation and cooperation advocated by DOE:

The Secretary shall consult and cooperate with the Governor and legislature of such State and the governing body of any affected Indian tribe *in an effort to resolve the concerns of such State and Indian tribe* regarding the public health and safety, environmental and economic impacts of any such repository. In carrying out his duties under this subtitle, the Secretary shall take such concerns into account to the *maximum extent feasible* (emphasis added).[15]

Specifications for the consultation and cooperation agreement, as outlined in the act, included: (1) terms and procedures by which host states could comment on the various impacts associated with a repository; (2) how DOE would respond to these comments; (3) how the host states would apply for impact mitigation funds; (4) procedures for information sharing and notification on key decisions; and (5) prenotification of high-level waste shipments to a repository. After identifying sites for characterization, the secretary was to "seek to enter into a binding written agreement" with these "host states," although no sanctions were mentioned if DOE failed to produce a consultation and cooperation agreement.[16] The act also included allowances for states with potential sites to conduct their own "reasonable independent monitoring and testing of activities on the repository site." These activities could check the scientific credibility and reliability of the DOE's science and engineering, but their significance was diminished by the condition that "such monitoring and testing shall not unreasonably interfere with or delay on-site activities."[17] Thus, DOE was under no mandate to delay the development of a repository even if state reviewers believed they had discovered, for example, geologic information that raised questions about the integrity of the site. (Later, the state of Nevada filed a lawsuit to force its right to engage in independent monitoring and testing, something which DOE saw as an unnecessary duplication of effort.)

Under the NWPA, a consultation and cooperation agreement between the DOE and host states was to specify procedures for "resolving objections of a State and affected Indian tribe at any stage of the planning, siting, development, construction, operation, or closure of such a facility within such State through negotiation, arbitration *or other appropriate mechanisms"* (emphasis added).[18] Similar to other language in the NWPA, it stops short of specifically requiring mediation or arbitration procedures which could enhance the negotiating position and power of the states. Absent a requirement that the states be directly involved in decision making, the act's definition of DOE's responsibilities emphasized public relations functions—informing the states and public, and considering their comments to the "maximum extent feasible." Andrea Dravo, a member of Representative Udall's staff when the act was passed, observed that states were granted an increased ability "to reject siting of a repository through building a technical case, or through inadequate procedural activity by the Department of Energy." However, she also noted that, "The first-round states have been largely shut out of the political process. . . . They are almost totally bound to technical and judicial leverage."[19] Proposals from the National Governors Association and members of Congress to use the Interagency Review Group or a state planning council as forums for broadening participation in decision making, especially at such key points as the drafting of the mission plan, were never incorporated into federal policy. In sum, the NWPA allowed only rudimentary forms of state and local involvement in the site selection process and in the entire high-level waste management program. First-round states, the public, and most other federal agencies, were limited to functioning as reviewers and commentators—not as participants actively shaping the goals and course of the repository program.

At the same time that requests for new institutions and forms of participation were denied, the act shored up the authority and discretion of the nuclear establishment. According to the NWPA, "the provisions of this Act shall not apply to any atomic energy defense activity or to any facility used in connection with any such activity."[20] Thus, facilities exclusively used for storing military waste would not come under the control of OCRWM. The Department of Energy's nuclear weapons facilities, operated by contractors such as

Du Pont and Rockwell, had already produced millions of gallons of liquid high-level waste. The Savannah River facility alone held thirty-three million gallons of high-level waste in the forms of liquids, salts, and sludge. Charges of improper management, cover-ups of leaks, and illegal releases of radioactivity were regularly leveled at this and other DOE weapons facilities. Supporters of military programs, who opposed NRC or any civilian regulation of weapons facilities, were successful in excluding military waste facilities from the NWPA. This kept the nation's first military waste repository—the Waste Isolation Pilot Plant in New Mexico—outside the authority of OCRWM and not subject to site selection, performance, and other requirements specified in the NWPA. Demands that military waste be regulated and bound by the same health and environmental standards that applied to its civilian counterparts went unanswered. Complaints about incompetent management of military high-level waste were unaddressed even though the act permitted the repository to hold military high-level waste.

More aggressive regulation of nuclear waste transport was also ruled out: "Nothing in this Act shall be construed to affect Federal, State, or local laws pertaining to the transportation of spent nuclear fuel or high-level radioactive waste."[21] The effect here was tacit approval of a U.S. Department of Transportation policy, supported by the DOE, of challenging state and local authority to regulate nuclear waste shipments. Thus, two additional sources of public distrust in DOE management—military waste and waste transportation—were left intact. Even though some members of Congress, the Office of Technology Assessment, and others claimed that restoring public trust and the credibility of federal institutions should be cardinal features of a nuclear waste policy act, the means for achieving this were never provided. The NWPA did nothing to create an independent organization which the public could trust to develop a credible and legitimate site selection and repository program. Apart from what it offered to site states, the NWPA granted no powers or rights of participation to federal, state, or civil institutions outside the nuclear establishment. Absent new institutions and sources of credibility and legitimacy, challenges to the authority claimed by DOE and other members of the nuclear establishment would continue.

However, in 1982 the institutional alternatives offered by the Reagan administration held even less appeal than those that already existed. Concurrent with its push for closure of the high-level waste issue, the Reagan administration made good on a campaign promise and introduced legislation (S.2562) to dismantle DOE and place a major portion of its budget and functions within the Commerce Department. The proposal was significant in that it suggested the possibility of an institution even more likely to favor the nuclear industry at the expense of environmental, health, and safety concerns. By introducing an alternative even worse than the present situation, the administration forced critics of the nuclear establishment into a position of having to defend the comparative desirability of current institutions, thereby distracting from efforts to introduce more radical proposals for reorganization. Opposition to reorganization also included the congressional armed services committees concerned with civilian control over DOE's weapons programs. In a Congress weary from proposals to reorganize federal energy agencies (President Carter had only reorganized ERDA into DOE in 1977–78), the legislation received only lukewarm support from some Republicans and outright opposition from the majority. The Congressional Budget Office concluded that it was unlikely to produce significant savings and feared that without DOE there would be no coordinated response to a disruption of foreign oil supplies.

Closing the Door on Institutional and Technological Innovation

For years the rhetoric, logic and ideology of the nuclear establishment reinforced commonly held perspectives on en-

ergy production and nuclear waste disposal. By retaining
DOE as the agency directing waste management activities
under the NWPA, Congress not only bought into the agency's technical and institutional orientation but also closed
the door on future institutional and technological innovation. While the need to explore alternative institutional
structures was indeed recognized in the NWPA, realistically,
the possibilities for institutional change were foregone
when the secretary of energy was assigned the responsibility
to "undertake a study with respect to alternative approaches
to managing the construction and operation of all civilian
radioactive waste management facilities."[22] The secretary
could construe this section of the act to mean that financial
alternatives, rather than new forms of management and organization, should be emphasized in the study. Nevertheless, the panel that was established to study alternative
means of financing and management (the AMFM panel)
went so far as to recommend that a public corporation take
over site selection, characterization, and repository operations. As a public corporation, a board of directors nomi-
nated by the president and confirmed by the Senate would
select a chief who, in turn, would appoint an advisory council responsible for recommending site selection procedures
and specific sites to the board. The board of directors and
the advisory council would include representatives of federal agencies, environmental groups, and the utilities. Proponents of this option cited the advantages as improved public involvement and confidence in the site selection process,
greater institutional responsiveness to environmental regulations, and a more diverse set of interests represented in
the decision-making bodies affecting the conduct and direction of the repository program. These were the same sort
of advantages that pre-NWPA proposals for institutional reorganization had attempted to capture. Of course, a major
unresolved issue was how to accomplish the transition from
DOE—already in the process of implementing the NWPA—
to a new waste management authority. The panel noted
that—even considering the serious defects in the manage-

ment structure of the OCRWM, the sordid history of its predecessor organizations, the "absence of credibility," and the panel's overwhelming support for institutional forms other than the OCRWM—"it may be difficult to effect any legislative changes in the NWPA."[23]

This suggestion of changes in DOE's management of the repository program came at a time when states and citizen groups were bombarded with thousands of pages of technical data and the legal implications of DOE's environmental assessments. The new repository program was underway and few took the AMFM panel seriously. In any event, implementing a change of institutions at this point would require reopening the NWPA, an alternative opposed by DOE, the utilities, and environmental groups. Environmental groups feared that a public corporation would be less responsive and accessible than a federal agency, and that it would be more concerned with public relations than with scientific integrity and credibility.[24] Potential host states feared that in response to their continued opposition to the DOE program, Congress would reopen the NWPA and eliminate the role of the states.[25] Aside from these concerns, institutional reconstruction was futile in an environment in which the AMFM panel made its recommendations to the secretary of energy who was unlikely to advocate that a "superboard" oversee DOE activities. From the outset, the AMFM study mandated by Congress was unlikely to effect major changes in institutional design, organization, or procedure.

Similar constraints extended into technical aspects of nuclear waste disposal. Research on alternative sites, long-term (though not necessarily permanent) storage, or technologies for reducing waste generation were curtailed or terminated when deep geologic disposal became the legislatively sanctioned solution. This occurred in spite of section 222 of the NWPA, titled "Research on Alternatives For the Permanent Disposal of High-Level Radioactive Waste," which called on the secretary to examine various waste disposal options and to investigate "alternative means and technologies for the permanent disposal of high-level radio-

active waste."[26] Concerned about achieving near-term progress toward a solution, many in Congress supported expediting a solution by relying on existing data and already
studied sites. Despite the advice of scientists skeptical of
repository technologies, key members of Congress came to
the conclusion that deep geologic disposal was the best option.[27]

Once innovation was restricted, it was inevitable that
federally sponsored research would be converted into technical demonstration. Test results would be used to confirm
decisions, and criteria would be adapted to decisions and
the "marketplace of ideas" would remain oligarchic. Research in other technological directions was impeded by an
overriding emphasis on permanent disposal. One example
was a small (less than five million dollars) DOE program—
referred to as the "extended burnup program"—which
examined the possibility of reducing spent fuel generated
by commercial reactors. The General Accounting Office
(GAO), in a letter to Representative Richard Ottinger, chair
of the Subcommittee on Energy Conservation and Power,
stated, "the administration in recent years has been trying
to phase out the extended fuel burnup program. . . . For
the last two fiscal years the administration has viewed the
extended burnup program as one that should be left to the
private sector."[28] Yet according to the GAO's assessment,
"Extended burnup of nuclear fuel provides a potential opportunity to reduce future spent fuel inventories and the requirements for government facilities and services to accommodate spent fuel. While the magnitude and timing of
reductions are uncertain, DOE has found that up to a 40
percent annual reduction in the rate of spent fuel generation
is possible."[29] The GAO observed that the potential benefits
of the program were being dismissed in the new administration's emphasis on "high risk, long term research programs."[30] In the push to settle on a final, permanent solution, Congress overlooked technologies which could reduce
the magnitude of the problem. Later, in 1986, DOE recognized that, with extended fuel burnup practices and slower

growth in the demand for nuclear-generated electricity, the projected inventory of high-level waste was far below the estimates offered when the NWPA was being debated. For example, in 1980 the DOE estimated an inventory of between sixty and seventy thousand tons of spent fuel by the year 2000 and as much as 203,000 tons by 2020. In 1986 the estimated inventory for the year 2000 was reduced to forty-one thousand tons. In the NWPA, Congress limited the capacity of the first repository to seventy thousand tons of high-level waste, which would have triggered development of a second repository under estimates of waste inventories prevailing at the time the act was passed. However, 1986 estimates showed that the inventory of spent fuel wouldn't reach seventy thousand tons until after the turn of the century, effectively eliminating the need for a second, eastern repository but also wrecking the commitment to a regionally equitable solution.[31]

The door to technological innovation closed tighter as section 211 of the NWPA authorized the secretary to provide for the construction and operation of a test and evaluation facility. In spite of its name, it was not defined as a facility for independent experiments that would test the scientific validity of proposed waste disposal technologies. In the language of section 211, it would "provide an integrated *demonstration* of the technology for deep geologic disposal" (emphasis added). This subtle but important difference continued to uphold the nuclear establishment's by now familiar assertion that the technology of deep geologic disposal was well established and all that was missing—and needed to address the critics of nuclear power—was an operational demonstration facility. Utilities and other nuclear industries had long sought a federal experimental facility to prove their point. Such proposals reached back to the days when proponents of a test and evaluation facility saw it as the answer to critics who charged that construction of new nuclear power plants should be prohibited until the nuclear waste disposal problem was solved. Similarly, in the first years of the Reagan administration, DOE spokesman Sheldon Meyers explained that the administration was calling for "the construction of a test and evaluation facility to *demonstrate* our capability to package, transport, and dispose of high-level radioactive waste" (emphasis added).[32] But the performance of a repository is necessarily theoretical because there is no way to unequivocally demonstrate that it will isolate radionuclides for a hundred, a thousand, or ten thousand years. "Proof" becomes a matter of establishing a consensus on the reliability and credibility of the models, data, and scientists employed to project repository performance. The Nuclear Waste Policy Act did not require a test and evaluation facility managed or staffed as an independent scientific body. The DOE was left to determine whether or not such a facility was even needed and how it would be used. Under DOE management, a test and evaluation facility could not furnish independent checks and assessments of repository performance. The agency gave the facility insignificant notice in the Mission Plan, defining its purpose in terms of engineering design work and testing hardware and procedures for repository operations.[33] As long as it was associated with an agency suffering serious credibility problems that had set out to *prove* (not test) the technology, a test and evaluation facility could never provide a believable evaluation of repository performance or reliability. Similarly, site characterization was not defined in the NWPA as a stepwise process in which research results would be periodically reviewed by an outside body, inadequacies remedied, and conflicts resolved before undertaking the next stage. Once site characterization was initiated no checkpoints were required by the NWPA. The secretary of energy was only instructed to report once a year to the NRC, the states, and Congress on the nature and extent of characterization activities and the information developed. To terminate site characterization activities, the secretary had only to inform Congress and the states of the decision and the reasons for it.[34]

Closure on innovation also followed a congressional decision to define the best approach to site screening and disposal, rather than defining a process for developing such an approach. Critical of DOE's emphasis on host rock formations, less sanguine members of the scientific community proposed alternative approaches for screening and selecting repository sites. For many scientists, there was nothing inherently desirable about one geologic formation over another because the ability of the repository to isolate radionuclides from the biosphere depended more on hydrologic conditions than on properties of the host rock. The U.S. Geological Survey suggested that the emphasis on rock types, rather than repository performance under different geohydrological conditions, was a simplistic approach to site selection and that the DOE had overlooked potentially suitable sites in the Great Basin. Nevertheless, the NWPA stated, "Such guidelines shall specify detailed geologic considerations that shall be primary criteria for the selection of sites in various geologic media."[35] Within two years after passage of the NWPA, DOE was preparing environmental assessments on nine potentially acceptable repository sites in Washington, Nevada, Utah, Texas, Louisiana, and Mississippi. The NWPA reduced the impact of outside research by organizations such as the USGS because its timetable for repository development allowed no room to redirect or expand the screening of potentially acceptable sites. The confines of relevant comment regarding site selection had already been established. Given the schedule laid out by the act, those who called for a new site screening program based on newly developed site-selection criteria were said to be "unrealistic."

From this point of view, the significance of the NWPA was not in what it had established but what it had left out. The NWPA exemplified how political bodies screen out some, and give political significance to other, sources of scientific knowledge. When the scientific evidence is uncertain or contradictory and offers no clear choice, legislatures can make a choice and leave others to create the applied science of implementation. Failing to create an environment favorable to independent testing and research, and the development of new approaches to site selection—apart from the nuclear establishment, Congress as much as said that the critics had had their say. Enough was known. Let the DOE go to work.

State Support and Subsidies for Nuclear Energy

By the time the 1982 NWPA was near passage private funding of the repository program was a nonissue. Electric utilities supported a special tax to finance a federally managed repository with the expectation that it would free the program from the uncertainties of federal funding. Because the cost of the repository would be simply passed on to its customers, the utilities were taking a cheap position— especially since other types of state support for commercial nuclear power were left untouched. Preemptory, federal regulation of materials transport and a cap on financial liability in the event of nuclear accidents continued. The Department of Energy continued to fund a network of national research and development facilities dedicated to reactor development.

Threats that nuclear power plants would be forced to close unless the federal government procured away-from-reactor storage were widely discredited by the time the 1982 NWPA was passed. Nevertheless, industry continued to oppose any legislation which did not include provisions for federal interim storage of spent fuel pending completion of a repository.[36] Congressional delegations from states dependent on electricity from reactors that faced storage problems pressed for federal interim storage. The Department of Energy supported congressional authorization of such a facility as a safety valve in the event that antinuclear groups successfully challenged other storage options proposed by nuclear utilities.[37]

The 1982, the NWPA established that reactor owners have the "primary responsibility for providing interim storage . . . by maximizing, *to the extent practical*, the effective use of existing storage facilities at the site of each civilian nuclear power reactor, and by adding new on-site storage capacity in a timely manner *where practical"* (emphasis added).[38] To the relief of utilities with near-term storage problems, DOE had the responsibility for "encouraging and expediting" the use of reactor sites and facilities and would also provide "not more than 1900 metric tons of capacity for interim storage of spent nuclear fuel for civilian power reactors that cannot *reasonably* provide adequate storage capacity at the sites of such reactors when needed to assure the continued, orderly operation of such reactors" (emphasis added).[39] Definitions of what constituted "reasonable" or "practical" were left to the DOE. Attempts to move the decision from DOE to the NRC were unsuccessful. The Nuclear Regulatory Commission claimed that its mandate was to protect public health and safety and that interim storage was a political-economic question not a safety issue. Thus, a federal facility for the temporary storage of spent fuel (be it an interim or monitored retrievable storage facility) became a recognized aspect of the nuclear waste program.

Under the NWPA, the DOE was permitted to continue its policy of helping utilities develop and license reactor-site storage. For example, Virginia Electric Power Company was running out of storage even as the NWPA was being debated. Conveniently, DOE took spent fuel from Virginia Power and shipped it to the Idaho National Engineering Lab for experimental purposes. By 1986 DOE had contributed about twenty-five percent, or eight million dollars, to the cost of developing a storage facility for Virginia Power in which spent fuel was stored in casks above ground and adjacent to a power plant.[40]

Following commencement of an operational repository, DOE was required to take title to high-level waste or spent fuel upon the request of its generator or owner; and, "in return for paying into the payment of fees established by this section, the Secretary, beginning not later than January 31, 1998, will dispose of the high-level radioactive waste or spent nuclear fuel involved."[41] Secretary of Energy Donald Hodel interpreted this to mean that he had authority to enter into contracts with utilities to accept spent fuel in 1998 whether or not a repository was in operation. Hodel observed that this would assist the utilities by permitting them to plan "their projected waste disposal needs with confidence and certainty."[42] The imperative of waste acceptance contracts could also be used to force repository development, the creation of a monitored retrievable storage facility, or to limit site investigations and public comment periods. In any event, utilities could look forward to a specific date when they could relinquish responsibility for handling spent fuel rods.

Divorced from a comprehensive assessment of the nuclear fuel cycle, the NWPA placed no limits on continued DOE efforts to fix the problems of the nuclear industry and to assist in the licensing of additional sources of nuclear waste—that is, nuclear power plants. A DOE task force, organized two years after passage of the NWPA, was handed the job of intervening in state regulatory agency proceedings to work for the completion of nuclear power plants, particularly troubled plants such as Diablo Canyon, Shoreham, and Seabrook. Continuing the agency's role as the promoter of nuclear power, Secretary Hodel testified on behalf of the Grand Gulf II plant and contacted New York governor Mario Cuomo about that state's opposition to the Shoreham plant. The DOE task force also sought "more reasonable" NRC regulations favorable to the industry and aided utilities in putting together their licensing petitions.[43]

In the aftermath of the Three Mile Island accident, proposals for omnibus federal nuclear legislation would have uncapped, or at least raised, utilities' liability for power plant accidents under the 1957 Price-Anderson Act. By its silence, the NWPA continued to protect utilities from the exorbitant costs of a power plant accident and left unaddressed state/local concerns about whether these liability limits applied to repository operations and nuclear waste transportation. Uncertainties remained about whether the Nuclear Waste Fund could be used to pay accident damage claims or for the clean-up of transportation or repository accidents. The NWPA was silent on Swedish-type recommendations which would have phased out nuclear power and on proposals, such as one offered by Senator Gary Hart (Colorado) that would have limited the growth of spent fuel inventories by prohibiting the NRC from issuing new licenses for nuclear power plants until a repository was operational. Thus, nuclear utilities were shielded from another threat to their capital investments and future earnings.

The 1982 NWPA also left intact a controversial federal policy of preempting local regulation and supervising nuclear waste shipments—already a source of serious federal-state conflicts. Thousands of shipments per year would be commonplace once a repository was operational. In an otherwise negligible discussion of transportation issues, the major concern expressed in the NWPA was that: "The Secretary, in providing for the transportation of spent nuclear fuel under this Act, shall utilize by contract private industry to the fullest extent possible in each aspect of such transportation."[44] Utilities were successful in convincing the federal government to take title to the spent fuel at the reactor site and assume responsibility for its transport to a repository. It would save them from the notoriety involved in dealing with state and local opposition along waste transport corridors.[45]

But even though the NWPA left the transport of spent fuel to private rail and truck carriers, the act included no provisions for addressing or funding improvements in federal-state-local coordination, monitoring, or emergency preparedness in corridor states. Congressional debate and numerous lawsuits on the inadequacy of the nuclear waste transportation system had been in progress since at least 1977.[46] Critics pointed to the fact that nuclear waste vehicle drivers were not specially licensed or trained. By holding multiple driver licenses, transporters could avoid suspensions due to safety violations. Federal inspections of carriers and cargoes were infrequent; when inspected by state officials, vehicles were often found without safety equipment—such as effective brakes. Enforcement of transportation regulations was difficult.[47] Nonetheless, the NWPA specifically stated that the act was not intended to affect existing federal, state, or local laws pertaining to radioactive waste transportation.[48] The U.S. Department of Transportation was free to continue its opposition to state/local regulation of nuclear waste shipments, even to the point of challenging local ordinances which prohibited transporting nuclear materials through city centers during rush hours. The U.S. Department of Transportation and DOE maintained that, because hauling spent fuel was less dangerous than hauling other hazardous materials, the existing level of regulation was sufficient. To reinforce its position that the transport of nuclear materials was safe, members of the nuclear establishment pointed to studies by the Nuclear Regulatory Commission that estimated that a significant release of radioactive materials in a transportation accident was virtually impossible. Thus, they concluded that state and local concerns about catastrophic transportation accidents and the need for expensive monitoring equipment and emergency preparedness networks were unfounded. Members of the nuclear establishment assumed that any accidents that did occur would be minor and could be handled by existing local emergency crews in consultation with federal agencies. The NWPA left open the possibility that local communities would bear the cost of responding to a radiological emergency and then try to collect damages from private haulers. In addition, it outlined no specific requirements for notifying state officials when waste shipments would enter a state; however, these procedures could be specified in the cooperative agreement that DOE would negotiate with site states.[49] Minimal prenotification requirements could make emergency preparedness more difficult for corridor states, but also reduced the likelihood of opposition groups demonstrating against nuclear shipments. Such demonstrations had already attracted the national media in Colorado and Wisconsin. Attempts to circumscribe corridor states' involvement were evident in attempts to limit participation to site states. The Department of Energy rationalized that corridor states would not suffer significant impacts and that including them would make the high-level waste program unmanageable.[50]

Another significant support for the nuclear industry that the NWPA left intact was a network of federal research and development facilities dedicated to expanding the production and use of nuclear power. Despite obvious problems with the management of high-level waste, nuclear energy continued to receive favorable treatment in the competition for research dollars. In budget discussions held concurrently with the debate over the NWPA, energy research and development was slated for $2.2 billion, of which sixty-seven percent would be for nuclear fission and fusion programs. At the same time, the Reagan administration proposed an eighty percent cut in solar and other renewable energy projects and suggested only thirteen million dollars for conservation research. The administration defended the cuts, saying, "The free market will determine the development and introduction rates of solar technologies consistent with their economic potential."[51] The NWPA presented no challenge to the nuclear establishment's view that continued federal sponsorship of nuclear energy was both necessary and desirable. After passage of the Nuclear Waste Policy Act, the DOE continued to justify the funding of nuclear research as necessary to assure domestic energy supplies and to maintain the competitive position of U.S. nuclear industries in international markets.[52] Utilities continued to resist demands that they pay an increased share of the cost of research on nuclear waste and other nuclear technologies.[53]

The 1982 Nuclear Waste Policy Act tackled the problem of high-level nuclear waste disposal by detaching it from issues such as the continued use of nuclear power. However, by adopting a narrow definition of the problem and its solution, the 1982 Nuclear Waste Policy Act failed to address key questions about the safety and reliability of nuclear technologies. By deferring to the demands of the nuclear establishment, it avoided building a basis for institutional and technological innovation. The act allowed DOE to con-

tinue promoting nuclear power and providing services to
the nuclear industry at the same time people were question-
ing the ability of DOE to identify a safe site for a repository
which would protect public health and safety. While the
NWPA introduced some changes, it failed to redirect nuclear
waste management in ways which would have addressed
important questions about the legitimacy and credibility of
agencies responsible for implementing and overseeing the
high-level waste management program.

CHAPTER 6

TECHNOLOGICAL
SOLUTIONS

WRITING IN a 1972 issue of *Science*, Alvin Weinberg, then director of Oak Ridge National Laboratory, suggested that the development of nuclear power placed unique and unprecedented demands on existing institutions to the point that it was necessary to consider the development of new institutions. The commitment to nuclear energy created implications that would last longer than any known government or institution. To Weinberg, the need for long-term supervision of radioactive materials suggested the creation of a nuclear priesthood—a permanent cadre of experts who, existing apart from short-lived national governments, would oversee the use of nuclear materials and energy.[1] These experts would form an elite unresponsive to political whims and demands and dedicated to the wise management of a technology based on universal scientific principles.

The idea for a nuclear priesthood was revived in 1984 in a report completed for DOE by Battelle's Office of Nuclear Waste Isolation, *Communication Measures to Bridge Ten Millennia*.[2] The report addressed Weinberg's question of how to prevent tampering with radioactive waste that would remain lethal for a hundred centuries after its burial, longer than any known human institution. The report suggested creating a nuclear priesthood, composed of scientists and scholars, which would become the self-perpetuating guardian of a nuclear waste repository. Stonehenge-type monoliths would be erected and universal symbols used to warn intruders of the dangers therein. Only the priesthood would possess the true knowledge of what was buried at the site. Myths and curses would be created and perpetuated to keep the ignorant away. This concept of a nuclear priesthood assumed the superiority of expert knowledge and the need to separate decisions about technology from the public and politics. This is the same assumption that relegates decisions on nuclear technologies to a techno-elite and that assigns the spent fuel disposal problem to the technicians. This chapter assesses the emphasis on specific technical issues and solutions in the 1982 NWPA and evaluates the implications of this emphasis for the implementation of the repository program and the reassertion of the nuclear establishment.

Technical Issues: Reducing the Role of Politics

The NWPA incorporated assumptions about scientific knowledge and technical expertise similar to those underlying the proposals for a nuclear priesthood. Removing politics from the business of waste disposal was a common theme throughout the debate on nuclear waste policy. Representative Morris Udall asserted that the site for a federal nuclear waste repository must be made free of politics.[3] In a retrospective assessment of the NWPA, Andrea Dravo, a Udall staff member at the time of the act's passage, observed that major goals of the NWPA were "to reduce the potential for litigation and procedural disruption," and, second, "to reduce the extent to which *politics* would govern the repository siting process, and to increase and protect the role *substance* would play in final decisions" (emphasis added).[4] According to Dravo, the act was successful on the second goal. In her opinion, it escalated the importance and sophistication of the technical debate in site selection. Specialized issues increased in importance when, through the Nuclear Waste Fund, site states received financial support to review the DOE's technical program, with the result that "the notion that the technical arena is central to the siting process has become an accepted part of the socio-political view, as is evidenced by declarations that the siting battle will be fought on technical grounds by many of the same elected officials who identify politics as a major siting determinant."[5] As used by Dravo, *politics* was a synonym for parochial interests.

Politics had dominated the site selection process. A number of powerful states effectively barred DOE and its predecessors from studying waste disposal sites within their boundaries. In 1976, Michigan's governor successfully challenged plans for exploration drilling in salt deposits. Plans for exploring salt sites in Ohio were dropped, at DOE's own admission, because of political pressures and a request from Senator John Glenn.[6] Less powerful states, such as Utah, were least successful in opposing DOE consideration of sites within their boundaries.[7] Throughout consideration of the 1982 NWPA, its sponsors beat back amendments intended to exclude specific states or jurisdictions from consideration for a repository. States west of the Mississippi unsuccessfully proposed a requirement that would bar transporting spent fuel more than five hundred miles from a reactor, supposedly in the interest of reducing transportation accidents. Under this limitation, potential sites, such as Hanford and the Nevada Test Site, could serve relatively few reactors, and a system of regional repositories would be required. Representatives of eastern states were successful in requiring that a repository could not be adjacent to a square mile area with a population of a thousand or more—effectively eliminating large portions of the eastern United States. The substitution of technical debate for politics was intended to limit this kind of self-serving legislation.

Two key expectations were imbedded in the desire to assert the primacy of technical decision making over politics. First, proponents of nuclear waste legislation such as Morris Udall hoped that technical decisions could be separated from geographic implications, but all technical decisions imply political-economic trade-offs that have specific, geographic implications. Determining what kind of bolt to use seems to be a straightforward, mundane technical decision best left to the engineer. However, the choice between a cost-saving, less durable bolt and a high tolerance, but expensive one can affect the potential for catastrophic failure of a hotel skyway or a nuclear power plant cooling system. Therefore, even this simple "technical" decision carries political implications: the specific geographic area and population exposed to a level of risk; the distribution of impacts and benefits associated with use of the cost-saving bolt instead of the more expensive one; and the social burden of responding to a catastrophe and treating its effects. Industry benefits from choosing the cost-saving bolt; however, the increased risk of injury or death is socialized. That is, the cost is distributed among the population at large. Complex decisions, such as which waste disposal technology or site to use, carry correspondingly complex and extensive political-economic implications. Thus, relegating a decision to the category of "technical" actually delegates political-economic power and responsibility.

The second expectation behind the emphasis on technological solutions was the belief that the separation of parochial interests from the decision-making process would result in a technically credible site selection program. Scientific objectivity would be substituted for partisan debate, and facts, rather than interests, would be the substance from which a solution was crafted. The goal was to develop a scientific basis for decisions and to produce consensus on technical questions. Some states expressed confidence in this approach, saying that if the data showed a location in their state as the best site, they would accept the decision.[8] However, the attempt to convert nuclear waste management into a scientific issue betrayed the hope that the resolution of technical uncertainties would clarify social choices and resolve value conflicts.[9] Studies of conflicts involving technologies with which humans have little experience (such as genetic engineering or nuclear waste repositories) find that, even when the issues are defined as technical, conflicts often cannot be resolved definitively on the basis of current scientific knowledge or technical practice. Tremendous uncertainties remain about what is the best site or the safest practice. Unlike the question of how to rebuild an automobile carburetor, consensus cannot necessarily be reached on the basis of existing technical standards or practice. Information on yet-to-be-implemented technologies may be inconsistent, contradictory, and ridden with ideological and geographical bias. For example, the fact that more studies were available to justify selecting salt domes in Mississippi rather than granite sites in New England said nothing about the relative suitability of those formations. When programs concentrate on improving the technical basis of decision making to the exclusion of conceptual and policy issues, scientific and social concerns become blurred. Focusing on the resolution of technical questions masks the conflicting values that underlie a dispute. Groups outside the nuclear establishment's technical community may see the resolution of conceptual, policy, and value issues as the central question. However, for the nuclear establishment, concerns about the independent replication and evaluation of scientific studies, or qualitative concerns about environment and culture or public participation, are not relevant technical questions.

An issue which was overlooked in the expectation that political decisions could be transformed into technical ones was that facts are produced by institutions. The facts of nuclear waste disposal are lodged within a network of federal agencies, contractors, and industries, which are no less partisan in the pursuit of their interests than a state or local government. In the case of high-level waste, the nuclear establishment dominated the means for generating the facts and technical basis for management decisions. It also had the reputation of withholding information which did not serve its interests. Scientific studies of Hanford workers' higher than normal cancer rates were suppressed for years after the data was turned over to "friendly" hands.[10] Scientists critical of AEC policies were fired or transferred, and data on the health effects of nuclear weapons testing in Utah and Nevada were suppressed for decades. Reports critical of reactor safety were suppressed until Congress demanded them. A U.S. General Accounting Office investigation into DOE studies at the Hanford site revealed that data on potential problems with the geohydrology of basalt had not been released as part of the environmental assessment.[11] The promotion of nuclear technologies by federal agencies and other members of the nuclear establishment had long ago compromised their ability to function as independent, disinterested producers of scientific fact.

The emphasis on technical over political decisions did not consider that science is actually fragmented among a variety of perspectives, with competing groups each claiming possession of fact and objectivity. University physicists and geologists who worked as part of the nuclear establishment endorsed the merits of existing waste disposal technologies, while their colleagues who worked for environmental and antinuclear groups criticized such endorsements as premature. For a new technology, technical and peer review processes were as likely to reveal the diversity of fact and opinion as they were to produce convergence and consensus on the facts.[12]

Reducing the Grounds for Political Opposition

The NWPA granted the state selected for a repository the right to give "a notice of disapproval" after the completion of site characterization studies, the final environmental impact statement, and the president's recommendation of a site. A state did not have an unqualified veto, and the NWPA required that a state's notice of disapproval be accompanied by a statement justifying why the site was unsuitable for a repository. Congress, however, could force a state to accept the repository by overriding its notice of disapproval. The NWPA identified the Nuclear Regulatory Commission as Congress' key source of information and comment on a state's notice of disapproval. Thus, congressional consideration of the state's case was likely to heavily weigh the technical reasons offered for the president's site recommendation.

Benveniste argues in *The Politics of Expertise* that, in situations involving contradictory evidence and challenges to institutionalized expertise, Congress is unlikely to contradict agency experts outright.[13] Barring glaring technical errors or flagrant violations of procedure, members of Congress would likely defer to the expertise of DOE and the recommendation of the president. To do otherwise risked reopening the site selection process and the possibility of having one's own state or district considered for a repository. Focused on site-specific and technical issues, the topics admitted to congressional debate on a state's veto were predefined. The demand for technical reasons justifying a state's veto limited the grounds for opposition to evidence of technical inadequacies at that, and only that, site. Questions of a better site or a flawed site selection process would be moot at that point.

Contrary to Dravo's assertion, the emphasis on technical issues and review did not necessarily enhance states' power. The Department of Energy was required to consider and respond to comments generated by states' technical review programs and to provide site states with money from the Nuclear Waste Fund to support such programs. Nevertheless, states' power to affect the waste management program was minimal. In the mold of "comment and consideration," participation became a symbolic act. Dissent could be tolerated because states did not participate in decision making anyway.[14] The equation of technical review with public participation raised the price of intervention for others. The development of competitive expertise was expensive, especially for citizen groups whose requests for funding were ignored by Congress and the DOE. Thus, with attention focusing on the technological aspects of high-level waste disposal, the power of technocrats to limit participation and, consequently, the topics of discussion was reasserted. This shift in focus reimposed order and limited the social complexity of the issues that would be addressed.

Benveniste suggests three ways for experts to shape policy and its implementation: by shaping the definition of the problem and its feasible solution; by controlling the readily available machinery of administrative action and policy implementation; and by directing unpopular decisions away from legislators to a remote, faceless bureaucracy engaged in engineering a technological solution. By focusing on the technical definitions of the nuclear waste problem and its solution, the NWPA permitted DOE to evade ideological conflicts over the goals of U.S. energy and nuclear policies. The NWPA also fragmented disputes into remote, narrowly defined technical issues, asserting the dominance of specialized language, information, and analysis.

The national park issue was a good example of how the application of technical analysis to a prominent, popular issue can fragment into obscure topics of debate. At the Davis Canyon and Lavender Canyon sites in Utah, the impact of site characterization on adjacent Canyonlands National Park was the important issue for national environmental groups, a majority of Utah residents, and many people in other parts of the country. While the environmental assessment of other sites received little media attention outside their home states, consideration of the Canyonlands "dump," was reported by the national media and distant big-city papers such as the *Boston Globe*. However, DOE and the contractors who actually wrote the environmental assessments looked to technical analysis to definitively demonstrate that site characterization constituted an insignificant impact on the national park. For example, impact was defined in terms of noise levels and visibility of the industrial development to park visitors. A large engineering company, whose major qualification was the construction of nuclear power plants, was hired to assess the significance of these impacts. In their assessment, the issue was converted into trivial discussion on whether painting buildings a neutral color would make them "invisible" to visitors, thus removing the impact. In the contractor's report, discussion focused on the measurement of ambient noise levels, definitions of the term *quiet,* and how many dynamite charges constituted a noise impact. Such analyses diverted attention from the larger issue—the unprecedented national cultural impact of locating a federal nuclear-industrial facility next to a national park. For most Americans, no amount of paint or measurements of noise levels could resolve the incompatibility of locating this facility next to a national park.[15] However, by cutting the issue into isolated, unrelated assessments of noise, air, water, and transportation impacts, DOE concluded that site characterization posed no significant impact. This finding was presented in the final environmental assessment and the Utah sites were deemed potentially suitable for a repository.

The professional culture of policy analysts reinforces this emphasis on technical issues and analysis. Commenting on the work of Harold Lasswell and his contributions to a broader contextual orientation to policy analysis, Douglas Torgerson wrote:

> In the context of advanced industrial society, there is a distinct and widely noted tendency for public policy analysis to become virtually absorbed in narrow, technical issues. This tendency has been especially noted in the case of efforts to rationalize the operations of the administrative state. . . . Under sway of a positivist logic of inquiry, analysis tends to be guided by an interest in calculating solutions for specific problems—ones which pertain, moreover, to strictly delimited frameworks.[16]

Torgerson goes on to note that professionalism may reinforce "the technocratic gulf" between expert and citizen, and align professionals with prevailing administrative institutions. My own experience in Utah's high-level nuclear waste office suggests this was indeed the case. State offices were largely confined to commenting on DOE technical reports, and little attention was paid to developing public education and involvement programs. Mushy issues such as cultural changes, quality of life, the national park, and environmental aesthetics—issues which nevertheless captured public and media attention—were downplayed and even dismissed as being an emotional and unreliable basis for contesting the selection of a Canyonlands site. The drive to implement a technology for disposing of high-level waste allowed the technical language, information, and analytic tools of the nuclear establishment to dominate policy discussions. Benveniste summarized this impact: "They [experts] provide a new language for policy-making and planning—a language that becomes the basis for limiting or enlarging the policy discourse . . . and also defines the time, place, and terrain of negotiations."[17]

In 1973 hearings on reactor safety staged by the Joint Committee on Atomic Energy, Del Sesto reports that pronuclear witnesses focused their testimony on highly complex technical data and risk assessments. Such data were offered as the way to resolve concerns about the continued use of nuclear technologies. Antinuclear witnesses, on the other hand, felt that the improvement of plant operations was not the primary issue. They raised questions about the burden that nuclear waste placed on future generations, the restriction of civil liberties necessary in a nuclear economy, the link between military and civilian nuclear programs, the value of preserving the "nuclear energy option," and the accountability of the AEC.[18] But to policy makers these big issues appeared intractable and complex. Dealing with such issues required wholescale rethinking of the nuclear cycle and federal energy policy. Congress did not want to open a comprehensive discussion of federal nuclear policy. Omnibus bills were quickly chopped apart, such as Morris Udall's H.R. 6390, introduced on 31 January 1980, which linked nuclear waste, safety, regulation, and liability with the continued use of nuclear power. Yet these same issues were carried into the implementation of the NWPA. Reasserting technical issues and the superiority of expert judgment denigrated the importance of arguments about values and reforming political institutions. Results such as fairness or equity are hard to measure, whereas the technicians of the nuclear establishment promised tangible results tied to schedules, site characterization, and construction. Critics of the nuclear establishment were dismissed as offering the uncertainty of a new site selection process, new institutions, and research.

People may never resolve their fundamental ideological differences over values and priorities—the big issues raised by critics of DOE policy. Possibly, they may come to agree on decision rules or at least the facts of the situation. An informational approach to conflict resolution assumes that if both sides share a common base of technical information, the chance of reaching consensus on specific actions to mitigate a problem is increased. The Department of Energy's approach to public participation stressed information exchange aimed at reaching consensus on technical standards and actions.[19] DOE's Mission Plan for the repository program continued this emphasis on informational approaches to conflict resolution.[20] This plan identified technical uncertainties that would be addressed and resolved during site characterization. Because DOE controlled the information base, it was in a position to control the nature of information-oriented negotiations, and to set the agenda of unresolved questions that would be topics for future discussion. The flow of information and discussion among experts within the nuclear establishment winnowed the "feasible" technical approaches and significant uncertainties.

After years of debate over what to do about nuclear waste, technology offered relief from continued ambiguity and uncertainty. With passage of the NWPA, difficult questions about values, institutional power and priorities could be put aside and attention could be focused on achieving consensus about more neatly defined technical issues. The network of engineers and planners within the nuclear establishment offered a ready-made way of restoring goal-directed, predictable administrative action. Technical facts appeared as a welcome substitute for the frustrations of political debate. The language of DOE's first draft Mission Plan was revealing in its assurance of the incontrovertible results to be produced by scientific studies. Rather than saying a particular study would address a research question or test a hypothesis, DOE maintained that a study would "confirm," "demonstrate," or "prove" the reliability of a geologic repository. DOE could focus on interactions between the scientists and engineers of the nuclear establishment who spoke the same language. In many cases, the topics for discussion were so specialized that there could be difficulty in finding individuals outside the nuclear establishment who were knowledgeable about the issue.

Researchers have found that information may have little to do with shifts in support for nuclear energy, suggesting that information-oriented approaches may do little to achieve common understanding and resolve high-level waste issues.[21] In such cases information is sorted to confirm political positions and ideologies. Looking at the same body of information on a new technology, opponents emphasize uncertainty, while proponents of a technology emphasize the investment already evident in the research. Because groups outside the nuclear establishment advocated different decision-making criteria, research assumptions, and sources of information, the nuclear establishment assumed that politics threatened technical judgment and substance. By defining politics as a threat to the technical resolution of the problem, the nuclear establishment was asserting that their science was superior to the science produced by outsiders.

If we assume that resorting to technical decision making defined a less partisan, less "political," more objective site selection process, then such a process should have been free from biases toward politically weaker jurisdictions. Technical expertise should have been used to defend the weaker jurisdictions from capricious political actions. However, as will be shown in chapter seven, this was never the case. Continued emphasis on technical issues worked against weaker interests and maintained the power of nuclear establishment members who controlled the machinery of technical justification.

Reasserting National Interests and Control

Both before and after passage of the NWPA, many members of Congress were concerned about the loss of public confidence due to their lack of resolve and inability to settle the waste management issue. Rather than converging upon a solution, the years of congressional debate seemed only to produce greater uncertainty about the hazards and reliability of deep geologic disposal. Members of Congress, especially those from districts where spent fuel and high-level waste were backyard problems, believed it was time to set goals and take decisive action. Rapid implementation and choice of a waste disposal technology was viewed as the way to reestablish federal leadership and national policy priorities and to restore public confidence in government's ability to solve problems. Thus, the NWPA became associated with a demonstration of congressional resolve. Dorothy Nelkin has noted that pressures to resolve a problem that is perceived as perennial or especially urgent leads to pressures to adopt any promised solution.[22] Not surprising, then, was the tendency for Congress to give wide discretion and authority to those capable of implementing that ready-made solution.

Permanent disposal in a geologic repository mined out of a salt, basalt, or tuff formation was not the only available option. Several European countries were experimenting with a combination of waste reprocessing and long-term, temporary surface storage (fifty years or more) while they investigated repositories in granite and other formations. Unlike the United States, Sweden viewed the ability of geologic formations to isolate high-level waste as essentially unpredictable. Instead of focusing solely on geologic containment (and the waste-isolating properties of a specific site), Sweden decided that any waste placed in a repository would be contained within multiple, long-lived (thousands of years or more) containers. In addition, that country had already decided to phase out nuclear power. While the United States continued to participate in some cooperative research projects with the Europeans, virtually all resources were committed to permanent, high-level waste disposal. The exit costs had grown too dear, and continued experimentation would not answer demands for decisive action to restore public faith in congressional leadership and federal control. When repository technology was defined as the "best available," future experimentation with alternative technologies was restricted.

In spite of warnings from the Office of Technology Assessment, Congress adopted a specific schedule for implementing a permanent solution in an attempt to eliminate future quarrels over the most desirable technology. Apart from offering more time to study the repository option, temporary storage left open the possibility of recycling nuclear materials—an advantage for groups interested in using them in the breeder reactor and weapons programs but opposed by groups concerned about nuclear proliferation. Problems at temporary storage facilities, such as the leaking Hanford tanks, gave temporary storage a bad reputation. Moreover, temporary storage meant that high-level waste was still a problem, only put on hold; therefore, it did not provide a rationale for lifting California's moratorium on new nuclear power plants. A temporary storage facility implied future debate and delayed resolution of the issue. Many viewed it as evidence of Congress' lack of resolve to solve the problem. For these reasons, temporary storage in a monitored retrievable storage facility (MRS) was never described in the NWPA as an alternative to a repository. A repository, which promised permanent and final disposal, matched the goals of the nuclear establishment members who wished to close years of political debate and uncertainty. They did not want to reconsider the problem in another few decades. One could argue that lack of legislative direction was responsible for past failures to address the high-level waste problem. However, after passage of the NWPA, blame for the lack of progress could now fall on nameless technical or bureaucratic factors.

It is interesting to note that, while the NWPA was meant to restore public confidence in Congress and the nuclear establishment, lack of confidence in existing and future institutions was used to justify permanent disposal in a geologic repository. With temporary storage, isolation of radionuclides from the biosphere depended upon conscientious management by future institutions. The poor record of nuclear management over the past thirty years left little reason to assume it would be more effective in the future. The questionable integrity of future institutions seemed to require a solution that was independent of human oversight. A repository was such a solution. Once its rooms were backfilled and the shafts sealed, its purpose could not be altered by political fiat nor could its inventory of nuclear materials be recovered by future bomb-makers. The integrity of a repository, unlike that of political institutions, could be defined in terms of geological time.

Technological progress on the waste disposal issue was tied to interpretations of the national interest. President Reagan's decision to "commingle" military and civilian waste storage supported arguments that a repository served more than just the parochial interests of the consumers and producers of nuclear power. Assuming that everyone benefited from programs which assured national security, some argued that, because the repository served the weapons program, it also served the national interest. Similarly, when dependable energy supplies were defined as a national security issue, anything that seemed to threaten the viability of nuclear energy—such as the lack of spent fuel storage—was also a threat to the national interest. Environmental groups argued that placing spent fuel in a repository that permanently isolated it from the biosphere was a more equitable solution because, unlike temporary storage, the risks and burden of nuclear power on future generations would be lessened. From the nuclear industry's point of view, a repository was advantageous because it answered the question of intergenerational equity, which environmental groups had often raised as a reason for opposing nuclear energy. Associating the repository with national interest was advantageous for those who wanted to characterize state/local opposition as self-serving and to limit local incursions into federal policy making. The association reinforced the opinion that the interests of the states were at odds with the national interest—even though states' concerns about inadequate protection of public health, safety and the environment could also be defended as in the public interest.

If the repository program served the common good, it did not achieve the goal set out by Russell Jim of the Yakima Nation—who called on all parts of the country to share the burden of nuclear waste disposal. Even if the benefits could be defined as universally shared, the costs could not, and under the NWPA DOE would determine how those costs would be defined and who would receive compensation.

Thus, the emphasis on technology and technocratic decision making were used to undermine state and local opposition. Technical substance was placed in opposition to "parochial" interests and politics, implying that opposition was motivated by self-interest whereas those in control of nuclear technologies served the public interest. If state and other groups wanted to affect the repository program, they were told to offer assistance in resolving the technical issues identified in DOE planning documents. In the end, the repository program established to implement the NWPA was the antithesis of innovation or a willingness to consider new approaches to nuclear waste management. The NWPA left the basic course of science and technology—set by the nuclear establishment decades before—unaltered. The expertise of the nuclear establishment was vindicated and the burden of proof was shifted to its challengers.

CHAPTER 7

▼

THE POLITICAL USES
OF LOCATION

A SOMNAMBULANT reporter walking into a DOE office in 1984 found scientists and planners single-mindedly engaged in selecting sites for a nuclear waste repository. Flow charts described a seemingly systematic process of screening sites—moving from the regional to the site-specific levels. Site selection criteria, which gave priority to long-term geologic stability, were used to nominate sites as potentially acceptable for a repository. Sites were classified using various ranking schemes and sensitivity analysis. The National Research Council's Board on Radioactive Waste Management judged DOE's method of ranking sites acceptable and found no bias in its application. DOE liked to quote the Board's findings in its press releases. Director of the Office of Civilian Radioactive Waste Management, Ben Rusche, said, "We are pleased and gratified that the Board has upheld the soundness of this technical step involved in the ranking process. . . . The findings of this prestigious body of experts should add confidence and credibility to this important national program."[1] Thousands of comments and technical documents had been considered; the country's largest engineering companies were behind the project. Lacking the historian's awareness, the reporter thought the site selection process looked impressive—a textbook case of systematic planning.

However, the site eventually selected for the repository—

Yucca Mountain, Nevada–was an indefensible choice if one viewed it as the best site selected from an inventory of all geologically suitable sites. Potentially acceptable sites had been identified long before the development of the site selection criteria required by the NWPA. Hanford, Washington, was not chosen as an acceptable site after an evaluation of all basalt sites; nor was Yucca Mountain, Nevada, chosen after an evaluation of all volcanic tuff sites. Consideration of other rock formations in the Midwest and eastern U.S. (figure 7) encountered bitter public opposition, was postponed, and eventually ended with the 1987 amendments to the NWPA.

DOE's list of potentially acceptable sites for the first repository was not the product of a detached, systematic inventory. The site selection criteria reviewed by the Nuclear Regulatory Commission and the National Academy were used to defend DOE's choice of sites, not to conduct a national inventory and screening of suitable sites. Site screening occurred within the institutional priorities and structure of DOE and its contractors. Attempts to rationalize the site

FIGURE 7. Regions considered for the second repository.

selection process by removing it from politics only pushed it further into the labyrinth of DOE offices where past agency actions could be defended and reaffirmed.

As observed over a decade earlier, the selection of a site for any controversial facility is made under threat; therefore, "we shall be assuming no longer the passive decision environment in which alternative choices are carefully and methodically studied until the best choice is made according to some set of defined criteria."[2] The choice of repository sites was made under threats from all sides. The nuclear industry demanded rapid implementation of an available technology; opponents demanded go-slow, open-ended, methodical study. Wolpert anticipated the political nature of the repository site when he stressed that location decisions are the product of stress and conflict within society, rather than the end result "of a dispassionate and considered selection of alternatives posited by the classical normative approaches or even the Simon scheme of bounded rationality."[3] With DOE in charge, choosing a location for a nuclear waste repository could never achieve the ideal of dispassionate, scientific objectivity. Geologists, hydrologists, or engineers could not settle the conflict as long as the credibility and legitimacy of the DOE and other members of the nuclear establishment remained issues. Under these conditions, the choice of a repository site was an exercise in political power—testing the ability of the nuclear establishment to assert its authority over its opponents. In contrast to idealistic models of the rational site selection process, the site chosen for the nation's first repository was an outcome of political, economic, and social forces which converged upon and defined the past and future uses of Yucca Mountain, Nevada.

Considered from the interests of political expediency, convergence on a location was a necessary part of the nuclear establishment's attempt to reassert its power and to prevent future challenges to that power. In chapter 6 we saw how technical definitions of the problem reasserted the influence of technocrats and engineers by narrowing the conflict to a small range of technical issues. Similarly, focusing on location effectively constricted and suppressed conflict.

Redefining a conflict as a locational one may restrict its magnitude by limiting the range of constituencies and issues that must be addressed. Reinterpreting a national issue as a site-specific one reduces the need to address larger questions of political procedure, organization, and equity. Thus, the NWPA's emphasis on locational issues affected the nature of the conflict, political participation, and the definition of costs and benefits. This chapter adds another layer to the theme of reassertion by showing how forcing controversy into the straightjacket of site-specific issues served the interests of the nuclear establishment.

Political-Economic Functions of Location

The federal government began considering salt sites for the disposal of spent fuel and other high-level waste in the 1950s. The 1960s and early 1970s saw numerous proposals for amassing waste in underground sites in Kansas, New Mexico, or in surface facilities. (These proposals have been described elsewhere.[4]) The problem of what to do with the waste was not inherently a problem of where to dump it. However, technologies such as transmutation—using chemical or nuclear processes to change waste into less hazardous substances—were unavailable and would require years of intensive research.[5] New medical, food processing, and other uses for significant amounts of nuclear waste products were also years away. The sale or donation of spent fuel to other countries' nuclear programs raised concerns about abetting nuclear proliferation—one of the reasons for President Carter's prohibition of reprocessing spent fuel. Reprocessing was commercially unprofitable and produced liquid waste which was difficult to dispose of. Recycling nuclear materials also threatened the once-through uranium fuel cycle which profited uranium mining and milling companies. The financial health of the DOE-owned uranium enrichment industry also depended upon a once-through fuel cycle. For these reasons, focusing on the "where to dump it" question served numerous interests, complemented existing technologies, and maintained the commercial nuclear fuel cycle.

Location also became associated with the ready-made, off-the-shelf solution. Unlike the invention of new waste disposal processes, there was nothing mysterious about finding a reasonably stable geologic formation. The data, field methodologies, and analytic tools for doing this were readily available and reliable. Implementing a locational solution could be done quickly, whenever it was needed. Unlike exotic solutions such as transmutation, it was not contingent upon the unpredictable course of basic research and invention. For members of the nuclear establishment, the availability of a site symbolized the certainty that the answer was in hand. The locational solution assumed that the federal government would maintain custody of nuclear materials. Waste disposal would not be subject to the risks of the marketplace which had plagued facilities such as the West Valley reprocessing plant.

Location implied political continuity and stability as well. The federal nuclear program, and the nuclear waste program in particular, had lurched through a decade of turbulent reorganizations. Furthermore, with the Reagan administrations' 1982 proposals for more reorganization, committing to a location symbolized the restoration of administrative stability. Changing public policy, corporate profitability, accidental releases of radioactivity, operational dependability, or maintenance would not have to be considered once the geologic repository was sealed.

Location is an accumulation of capital. For example, the Nevada Test Site and Hanford encompassed billions of dollars of capital investment. Capital was concentrated and fixed in the form of reactors, waste storage tanks, low-level waste burial sites, transportation links, offices, labs, and employee housing. These locations were hubs of the U.S. nuclear economy; therefore, it should not be surprising that they were eyed as hubs for the civilian waste management program, too. Even before passage of the NWPA, much of the equipment needed for site characterization activities had been purchased or was already available at these sites.[6] Public land, DOE field personnel, offices, contractors, communication, electrical, transportation and security networks—all requisite infrastructure for a repository—were there. The post-NWPA repository program tapped these resources and facilities, expanded opportunities for resident expertise and labor, and contributed to local economies. Once the sunk costs of developing the basic services at a location were paid, new users were attracted to take advantage of its agglomeration economies. This simple economic logic applies to city industrial parks as well as federal nuclear reservations. For groups that had to foot the bill for spent fuel disposal, adding these functions to an existing nuclear reservation was more efficient than developing a location from scratch. With the legacy of past capital investment, these locations reduced the costs of initiating a new waste management program.

It takes months, or even years, for a new house owner to learn where to shop, bank, get a car repaired, to find a reliable plumber, a good neighbor, and businesses to avoid. Thus, location becomes a form of social accumulation. Management organizations and hierarchies, data banks, budget, purchasing, and supply procedures all have to be worked out for a location. And so site-specific expertise evolves. A location reflects a unique body of information about itself and connections with other places. In some cases this information may be highly detailed and the result of a huge investment. For example, a vast data set on the geology of the Hanford site had accumulated for years (though some contested its reliability). The huge network of information and expertise centered on already-studied locations simplified and reduced the cost of planning and decision making. Networks of specialized engineering and scientific services were highly developed at a some locations, such as Hanford and southern Nevada. Unique facilities and services, such as radiological decontamination units, or hospital staffs familiar with radiological emergencies, were available.

A lack of local political institutions supportive of nuclear industries could create huge uncertainties for a developer, putting the project at the mercy of unpredictable, personalized forms of local decision making. Boomtown politics can be volatile—fueled by resentment toward outsiders, local elites mindful of their power, and a local culture protective of the status quo.[7] A desirable location for a repository embodied stable, supportive relationships between state, business, and civil groups that looked favorably upon the nuclear establishment.

Engineering may be the application of principles of physical sciences, but it parallels economics and business management in its attention to labor costs, production schedules, and government regulations. As long as efficiency and practicality are valued aspects of engineering, the advantages of grafting new uses on to existing nuclear reservations were attractive. Schemes for starting over with a novel, national site screening program eliminated the practical advantages of already-studied locations and the concentration of economic and institutional capital evident in places such as the Nevada Test Site. While a new site selection process could have more closely approximated the ideal of rational planning, such a proposal confronted the competing ideals of efficiency and practicality which favored the choice of federal nuclear reservations as prime candidates for the repository.

Failure of the opposition to initiate a new site screening program pointed to the difficulty of wrenching locational analysis from its economic and political contexts.[8] New processes for identifying new sites required new institutions. Existing locations preserved the hierarchy and distribution of power among existing institutions in society, favoring the sponsors of nuclear power and putting its critics at a disadvantage.

The Location of Public Acceptance

Location also defined a situation where public acceptability of a repository was likely to be higher. Of course this was never an explicitly recognized reason for choosing a site. The Department of Energy maintained that it had been studying various sites for over twenty years. Yet, in the end, the nine potentially acceptable sites that were chosen shared conditions favorable to political acceptance of a repository.

Many of the sites were in areas where nuclear industries already had made a substantial contribution to the local economy. The Davis and Lavender Canyon sites, in southeastern Utah, were in an area of intensive uranium mining that had once brought a period of prosperity to an otherwise remote and depressed ranching economy. The Nevada and Washington sites were on federal nuclear reservations which contributed millions of dollars to state and local economies. In 1979 Department of Energy spending in the state of Nevada (approximately $357 million) exceeded the Defense Department's ($334 million) and was the equivalent to nearly $450 per state resident. DOE outlays in Washington were $1.14 billion, or approximately $276 per state resident. But these figures hide the fact that the outlays were concentrated in the Pasco-Richland area—site of the Hanford Nuclear Reservation. In Texas, the Deaf Smith County site was about thirty-five miles from the Pantex plant in Amarillo—where U.S. nuclear weapons are assembled.

Many locals supported existing and additional nuclear industries. While opposition in small farming towns near the Texas site flourished, residents in larger nearby cities, such as Amarillo, did not actively oppose the dump. In *Blessed Assurance: At Home with the Bomb in Amarillo, Texas,* A. G. Mojitabai found that most Amarillo residents seemed to accept the Pantex nuclear weapons plant as part of the landscape. With a disconcerting equanimity, a local minister said, "It's kinda like knowing the Santa Fe Railroad is here. . . . You know it hauls things. You don't really pay attention to what percentage of cars are coal cars, or whatever."[9] A DOE official associated with the plant interpreted the lack of local workers' concern by saying, "Hell, people sleep on those things."

Local politicians at the Nevada, Washington, and Utah sites had long championed the repository as a desirable new industry. Southeastern Utah politicians and businessmen, recalling the prosperity of the uranium boom, championed the repository as a much needed economic development for the economically depressed area. Polls conducted by the city council of Monticello, Utah, claimed that sixty-three percent of the local residents favored site characterization testing, and sixty-one percent favored a nuclear repository, "if the testing proves the site safe."[10] County commissioners and a local city council criticized the governor's opposition to the repository, calling it insensitive to local economic conditions and charging that "Your office has repeatedly ignored the concerns of elected representatives of those people who are closest to the Gibson Dome site and focuses instead on politically popular environmental issues. Your statement exaggerates both the size of the potential storage site and its impact on Canyonlands National Park, thus misleading numerous voters in Northern Utah."[11] Similar expressions of support for a repository were made by politicians from small towns near the Nevada and Washington sites.

Another factor which favored political acceptability was the association of rural, low income populations with low levels of political activism. The Richton Dome and Cypress Creek sites in Perry County, Mississippi, were among the poorest areas of the United States. (The same was true for Louisiana's Vacherie Dome, but it was never really in the running.[12]) The Department of Energy promised Richton, Mississippi (population 1,100), an economic boom. Of the millions to be spent on characterizing the site, DOE noted that approximately thirty percent would go for wages and, "It is likely that some materials (such as fuel, concrete, small equipment, lumber and other building supplies) will be purchased locally. A part of the wages will be spent locally, including wages from indirect jobs generated by the project."[13] Local and state vulnerability was obvious. In a comparative study of Michigan and Mississippi's response to high-level nuclear waste siting, political scientist Susan Hansen found: "Historically, Mississippi has had low voting turnout among whites as well as blacks. . . . Mississippians have limited opportunities to vote on referenda, initiatives, or Constitutional amendments, and thus have fewer opportunities to express their opinions concerning nuclear power or nuclear waste."[14] Hansen's findings suggested that even if Mississippians objected to the import of "Yankee" wastes, the need for storage of military waste was likely to go unquestioned. Support for military programs was strong in Mississippi. In 1979, Department of Defense programs budgeted for the state equalled $557 per capita. Mississippi Senator John Stennis, chairman of the Senate Armed Services Committee until 1981, had opposed increased oversight and environmental regulation of DOE's military waste program.

Recognizing that Mississippians were politically ill prepared to deal with DOE, state officials sponsored training sessions to help local residents overcome their lack of political expertise and organizational skills. However, Hansen points out that Mississippi had already been on the losing end of many federal-state confrontations. Past failure bred an expectation of future failure among local officials. This contrasted with Michigan's successful opposition to DOE site investigations. Opposition in Mississippi was also hampered by the state's decentralized form of government. Hundreds of departments and bureaus comprised state government; little power was vested in the governor's office. The state had an arcane and complex constitution hundreds of pages long.

Other repository sites shared many of these same political and economic conditions. In 1980, the population of rural Nye County, Nevada, was 9,048, or 0.5 persons per square mile. The Department of Energy expected a local population increase of over 16,000 due to repository construction, of which at least 2,100 would become residents of Nye County and form a powerful new voting block. Southeastern Utah had similarly low incomes and population density. A repository in a place such as Nye County or southeastern Utah implied less aggressive environmental regulation and a reduced likelihood of attempts to regulate nuclear industries.

None of the states save Washington, had a history of nuclear referenda. In Mississippi environmental groups were almost absent. In Utah and Nevada small groups of environmentalists struggled against state and local governments inhospitable to government regulation, environmental preservation, and even national parks. Louisiana had notoriously ineffective environmental regulations, in part due to that state's receptivity to the oil and chemical industries; nor was Texas known for aggressive environmental regulations. Washington had a curious history. Former governor Dixie Lee Ray had been chairwoman of the AEC, and, in the 1970s, promoted the idea that nuclear waste should be viewed as a resource. Washington utilities had aggressive plans for expanding their nuclear generating capacity. Environmental activism, centered in the Puget Sound area, was much less visible in the sparsely populated eastern half of the state where the Hanford Nuclear Reservation was located.

Finally, political acceptability was increased because, with the exception of Texas, potentially acceptable sites were within states that commanded an insignificant role in presidential elections—unlike the states once considered for a second repository. The effectiveness of host states' congressional delegations to affect site selection is difficult to sort out. All delegations to some degree attempted to have their state excluded from consideration; but, typically, their stance on nuclear technologies was either ambiguous or openly supportive. In the years preceding passage of the NWPA, other states' delegations appeared to give higher priority, and dedicated more resources, to resisting DOE than did the first-round states. For example, Senator Henry M. Jackson of Washington, a ranking member of the Armed Services and Energy committees, was a strong supporter of nuclear power and the nuclear weapons program. He voted against a six-month moratorium on new construction permits for nuclear power plants following the Three Mile Island accident. His opposition killed a 1980 nuclear waste policy act, favored by many environmentalists, which would have required military waste to come under the same regulations applied to commercial nuclear waste.

No surprise was Washington Representative Sid Morrison's promotion of an amendment to exempt Hanford from health and environmental regulations—to assure that his district *was* considered for a repository. From 1971–81, Morrison's district, home of DOE's Hanford Nuclear Reservation, had been represented by "Atomic Mike" McCormick. Other Washington representatives, such as Al Swift, who represented a Seattle area district, were less enthusiastic about Morrison's proposal. Senator J. Bennett Johnston (Louisiana) had few supporters, and strong opposition from Representative Morris Udall and the Illinois, New York, and South Carolina delegations, in his battle to adopt a monitored retrievable storage (MRS) facility as the preferred solution. As a surface facility, an MRS facility could separate the solution to spent fuel storage from its dependence on local geological formations, such as Louisiana's salt domes. The NWPA included a weakened MRS provision which called for further study of this option and a recommendation from DOE on whether it should be built. (In 1987, Senator Johnston would rise again. This time he successfully championed amendments to the 1982 NWPA which effectively eliminated from consideration all sites for a repository except Yucca Mountain, Nevada.)

In terms of national politics, Texas was in the most powerful position of any site state delegation. But a review of congressional testimony and reports of the *Congressional Quarterly* found little evidence of vigorous opposition from the Texas delegation before passage of the NWPA. In fact, Texas Representative Phil Gramm pushed a bill, favored by the nuclear industry and the Reagan administration, intended to make it easier for the federal government to override a host state's veto. According to 1982 reports by *Congressional Quarterly*, Gramm and a bevy of DOE, administration, and nuclear industry lobbyists repeatedly beat back amendments by environmental lobbyists which would have strengthened state oversight of DOE's program. Opposition in Texas became more visible and vocal after DOE officials announced in 1983 that Texas held two potentially acceptable sites. The following year it was evident that politics were as much a part of the site selection process as ever.

According to a letter quoted in the *Dallas Morning News*, Secretary of Energy Hodel informed Gramm—a former Democrat, now turned Republican senatorial candidate:

> I do not think a nuclear repository will be built in a state such as Texas where there is powerful opposition from the people and from powerful representatives of the people such as Phil Gramm. . . . It would be in direct violation of the spirit and letter of the act of 1982 if during this process we failed to take into account the strong opposition of Congressman Phil Gramm to locating a nuclear waste repository in Texas.[15]

The announcement that Texas held one of three sites chosen for characterization was delayed until after the 1984 elections. However, allegations of election-year tampering with the site selection process may have assured that, to save face, DOE chose Texas as one of three sites for characterization.

Mississippi's Trent Lott, Republican representative of the district with the Richton site, was successful in including language in the 1982 NWPA which barred construction of a repository if its surface facility was adjacent to a square-mile area with a population of 1,000 or more. However, its effectiveness in protecting Mississippi was still uncertain since surface facilities could be moved to avoid the restriction. Some took comfort in the fact that a member of Senator John Stennis' family lived in Richton. But in 1982 the senator was eighty-one and lost his chairmanship in the Republican takeover of the Senate. Everyone expected he would not seek reelection. Ironically, Texas Senator John Tower replaced Stennis as chairman of the Senate Armed Services Committee but generally remained aloof from the debate on the 1982 NWPA.

Interestingly, a senator from a nonrepository site state, William Proxmire of Wisconsin, threatened a filibuster which would have killed the 1982 act. Earlier drafts of the NWPA required one house of Congress to sustain a state veto. The "state's right" amendment demanded by Proxmire, and at the last minute incorporated into the act, required that a state's veto stand unless both houses of Congress voted to override it.

Defining Constituencies

Political geographers have noted that modern politics is based on a system of geographically distinct constituencies and representatives.[16] Members of the U.S. Congress are not elected at large but by state and district constituencies who expect them to demonstrate responsiveness to local demands. Congress is not only an arena for debating national policy but is also an arena for political-geographic contests. Broad-based interest groups that transcend local boundaries must nevertheless work within such a system and build coalitions of geographic interests. For example, regardless of the national sympathies in its favor, a new national park traditionally is not legislated over the disapproval of the local congressional representative. While political-geographic views can be carried too far—for example, all politics can not be explained as parochial, pork barrel politics—national policies invariably produce geographically specific effects.

The Nuclear Waste Policy Act defined waste management in terms of specific locations and thus specified a set of jurisdictions within which future political interactions were supposed to occur. Identifying a repository site implied a limited set of interactions among federal, state, and local

jurisdictions. In cases of environmental impact, local jurisdictions were used to define affected residents in need of
compensation. One county could receive impact mitigation
funds while an adjacent one could not. Local residency became a necessary criterion for participation in schemes for
mitigation and compensation. Thus, transforming the problem from a national policy issue into a locational one nar-
rowed the range of institutions that could claim jurisdiction
over the issue. The focus on location limited the spatial
extent of interests that had to be accommodated. A sort of
spatial franchise was granted, with some constituencies receiving the right to participate in negotiations and privi-
leged consultation with DOE decision makers. Those outside these boundaries were denied equal status and access.

The political implications of location became obvious
when the NWPA identified the state as the key jurisdiction
with whom DOE would interact. States with sites were
granted the status of host states, which translated into the
right to receive payments from the Nuclear Waste Fund.
Their participation in the repository program, in the form
of nuclear waste offices and technical review groups, was
funded by DOE and their comments given special consideration by DOE planners. Only the governor and/or the legislature of host states could veto a final site selection. A host
state exclusively could request the president to halt the repository program until a state's inquiry was answered. And
only host states were identified as the party with whom
DOE was required to attempt to negotiate consultation and
cooperation agreements.[17] City, town, and county governments were granted no specific rights—nor were environmental or citizen groups. State governments were responsible for addressing and protecting local community interests.
State nuclear waste offices tried to force DOE to channel
all site-related communications through them. This was to
prevent DOE from bypassing the state and negotiating directly with local communities, especially rural towns, who
might be more receptive to offers of economic incentives.

State officials were also interested in controlling the allocation of funds received from DOE since the amount of money involved was significant. Between January 1983 and March 1985, DOE granted $1.7 million to Utah, $1.9 million to Mississippi, $2.9 million to Nevada, $2.7 million to Washington, and $0.6 million to Texas.[18] Identifying state government as the key point of contact simplified the institutional landscape for DOE. It could deal exclusively with state-level offices and ignore, when desirable, regional groups and local governments or citizen groups. This reduced the number of parties and clearance points required to make progress in the repository program. The NWPA did recognize the need for a regional approach to repository siting as a second, eastern repository was introduced to satisfy demands for a regionally equitable solution. However, the NWPA established no formal institutional mechanisms, such as a state planning council, to review regional issues. The site state approach left other states and organizations with little leverage by which to force a regional approach to environmental impact assessment. For example, the Texas site raised questions about contamination of the Ogallala aquifer, which extended into Nebraska and Kansas and supported a regional, agricultural economy. The Utah site was likely to spill boomtown effects into nearby towns in Colorado. Impacts associated with the transport of nuclear waste to the repository meant that nonhost states had an interest in the site selection decision. Even if Utah was essentially out of the running for the first repository after 1984, the transportation route to the Nevada site required spent fuel shipments to pass through the middle of Salt Lake City. Highway and rail corridors to southern Nevada bisect the elongated Salt Lake City metropolitan area, and nearly the entire population of the state lives within five miles of these transportation lines.

For Colorado and Nebraska, a repository in Utah or Nevada raised the possibility of thousands of shipments per year traveling through their jurisdictions. Colorado feared that trucking companies hauling spent fuel might cross the Continental Divide on Interstate 70. Sections of that highway had high accident rates, were closed for days in the winter, and crossed mountainous terrain that complicated emergency response. Added to this were concerns that routing and the safety of interstate truck traffic would be poorly monitored. The major rail route for spent fuel traveling across Colorado was not much better: it followed the Colorado river, a treacherous route in which a spent fuel cask could fall hundreds of feet, landing in the Southwest's most important river. Thus, states such as Colorado and Nebraska wanted regional transportation issues to be considered when selecting and developing a repository site.[19] The Department of Energy responded to these concerns by saying they would be considered in the environmental impact statement (EIS) which would accompanying a final recommendation for a site, scheduled for sometime in the 1990s.[20]

The zone of impact defined by DOE in its environmental assessments meant that only site-specific transportation issues—local railroad crossings, road upgrades, and the construction of rail spurs—were addressed. A number of reasons were offered to justify limiting the geographic extent of the environmental assessment and to dismiss regional transportation issues. DOE pointed to the safety record and lack of transportation-related fatalities. Because spent fuel casks were believed to be capable of withstanding the worst likely accident, highway safety and routing were viewed as largely irrelevant to site selection. According to DOE-sponsored risk assessments, public fears about a cask rupturing in a transportation accident and contaminating river systems were unfounded; the radiological effects of a cask moving through urban areas were negligible. Adding to radiological and emergency response capabilities of local jurisdictions along transportation routes could always be done later and were unlikely to be needed, in any event, because of the reliability of the spent fuel cask.[21] Socioeconomic impacts outside the immediate repository area were expected to be so diffuse as to be virtually unrecognizable. Distant cities' and states' concerns about a repository leaking radioactive waste into the Columbia or Colorado river systems

or the Ogallala aquifer were unfounded and would be ne-
gated by mine site engineering.[22] Confidence in the site's
integrity and repository program engineering meant there
was little justification for extending the equivalent of host
state status to states dependent on the Ogallala aquifer or
downstream states on the Colorado and Columbia rivers.
By emphasizing only host state involvement, the NWPA
drastically reduced the number of groups and jurisdictions
with which DOE was required to negotiate. DOE offered
little beyond this condition—to the relief of utilities who
feared that permitting broad participation would inflate the
cost of the program and only encourage citizen protest and
delays.

Early emphasis on location fragmented constituencies
that otherwise may have shared similar interests and agendas. Once possible locations were known, attempts to affect
policy and the repository program were divided among site
and nonsite, safe versus threatened states. At their worst,
relations among the site states degenerated to pointing the
finger at another state as having the best site.[23] Fragmenta-
tion was evident in state politics within Nevada, Utah, and
Washington as locals supportive of the repository battled
state officials critical of the repository. Even if a state office
was supposed to be the primary point of contact, DOE maintained no liaison offices in capitals of site states but did
maintain information offices in towns near the sites. Visits
by DOE personnel to state offices were rare. National groups
critical of the nuclear establishment were denied the status,
resources, and membership privileges granted to host states.
Apart from DOE's assurance that "your comments will be
considered," institutionalized mechanisms for involving
representatives of civic groups—who remained a major
source of opposition to the repository program—were
never required or created by the NWPA. The focus on specific locations hindered the development of forums for addressing broad-based issues and nationally significant opposition to the DOE program. It is no surprise, then, that
groups critical of the waste management activities of the

nuclear establishment took their campaign to the mass media and the courts.

The complex geography of waste management issues could wear down opposition and discourage the entrance of new groups critical of the nuclear establishment. Consider the multiple and unclear boundaries between the offices of DOE, its numerous contractors, and other federal agencies. (DOE even hired contractors to supervise its contractors.) Before they were terminated, repository activities in Texas were directed from the Office of Nuclear Waste Isolation and the Salt Repository Project Office in Columbus, Ohio, which was managed by Battelle National Labs for DOE. As originally planned, an office of Battelle (DOE's "integrating contractor") would be located in Hereford, close to the actual site. The project's field services contractor would be located in Amarillo. The construction contractor's office would be in established in Vega, along with a public information office. The main OCRWM public information office would continue to be in Washington, D.C., and the state's nuclear waste office was in Austin. DOE would be responsible for constructing and operating the repository. The Nuclear Regulatory Commission would oversee enforcing performance regulations and standards promulgated by the EPA. In addition, the NRC wanted to make sure that data collected during site characterization was conducted under a proper quality assurance plan. Such complexity made it difficult for civic groups and newcomers to identify targets for protest, the proper channels for placing requests, or the locus of administrative authority and discretion. Multiple locations complicated attempts to pinpoint institutional accountability. Complicated institutional hierarchies and procedures made it difficult to identify avenues for effective participation. Who was in charge of the waste management problem or responsible for addressing a specific issue—the OCRWM Washington, D.C., office? the Columbus office? the local project office? which contractor? which agency?

For each repository site, fragmented and unclear definitions of responsibility allowed problems to be deferred to other agencies and levels of authority.[24] Responsibility and culpability for problems could be shifted and confused. Inaccurate or inflammatory statements made by a contractor or DOE official could be blamed on poor interoffice coordination. Delays in responding to states' inquiries were blamed on the need to consult legal counsel or on problems within a contractor's office. Contractors delayed answering a state's request until they received clearance from DOE headquarters. The nuclear establishment's knowledge of its own complexity helped it control public access and participation. Protest could be channeled through complicated, time consuming, and ultimately discouraging bureaucratic procedures.[25] Sustained involvement in a program scheduled to last decades demanded an extraordinary commitment of time, resources, and concentrated attention that most citizen groups or States could not afford. As a result, active participation by civil groups remained concentrated in a few sophisticated environmental organizations such as the Natural Resources Defense Council and the Environmental Defense Fund. As Gaventa noted in his study of power, "Power serves to maintain [the] prevailing order of inequality not only through institutional barriers but also through the shaping of beliefs about the order's legitimacy or immutability."[26] Nevada residents, faced with challenging the mass and complexity of institutions comprising the nuclear establishment, often remarked that, whatever the state—and regardless of what the governor—said, Nevada was going to get stuck with the repository.

Manipulating Costs and Benefits

Externalities are defined as costs and benefits not accounted for in a transaction. They can be created and their geographic distribution affected by political action. State inattention to pollution control, for example, allows a manufacturer to avoid the cost of air pollution control equipment but forces the public to assume the environmental cost of air pollution. The geographic distribution of such a cost is referred to as an "externality field."[27] The power to influence or direct policy implementation can be expressed as the power to create and distribute positive and negative externalities. And the power to depoliticize or defer consideration of an issue—making it a nonissue—is the power to maintain the existing distribution of externalities.[28] The authority to implement the NWPA had similar implications for the recognition and geographical distribution of externalities.

First, by viewing nuclear waste disposal as a locational problem—and viewing opposition to a repository as a locational conflict—the range of estimated externalities was substantially reduced. Attention became focused on site-specific, rather than regional, externalities. In DOE's environmental assessments, analysis was limited to impacts on local services, the economy, and environmental quality.[29] Impact analysis was reduced to estimating how many fire hydrants, tennis courts, and new apartments would be needed and at what price. The social engineering necessary to remedy these kinds of problems was conveniently in hand, having been used to garner local support for large industrial projects elsewhere in the West. An available body of research and expertise, developed during the era of energy boom towns, could be tapped and its methodologies transferred to the repository situation. Housing costs, adequate police and fire protection, and schools, for example, were real concerns, as the residents of Gillette, Wyoming, and other western energy boom towns knew. However, these concerns could be identified, quantified, and accounted for; they presented an inherently solvable problem. Costs and compensation could be calculated, bargained over, and negotiated to settlement. Local issues could be made into nonissues. The conflict over site selection could be depoliticized if issues were kept compartmentalized and irrelevant to broader, outside constituencies. By concentrating on site-specific externalities, the waste management program focused on issues that were inherently solvable, unlike the greater uncertainties and threats involved in addressing demands for procedural and organizational reform.

The locational view assumed that conflict was dysfunctional, a source of social disruption, paralysis, and inefficiency rather than an indication of institutional or systemic deficiencies. Viewing opposition to the OCRWM program as a locational phenomenon held the obvious advantage of defining the conflict in terms that would reduce its scope, focus on location, and fragment the controversy into specific problems that could be objectively defined and eliminated by financial compensation or technical fixes.

In one case, however, localized externalities became nationally significant. Otherwise local issues, such as the impacts of road building and other site development activities, attracted national attention when they threatened the pristine Canyonlands National Park. Blasting of mine shafts, heavy truck traffic, and fugitive dust—which at other sites would only be a nuisance for a few locals—suddenly became topics of national debate. Environmental groups pointed out that DOE was not concentrating the disadvantages of nuclear waste in some remote and ugly locale but was passing them on to all U.S. citizens who had an interest in protecting the beauty of their national parks. They demanded national hearings and public meetings. The media questioned DOE's judgment. How could it conceive of turning an area of great natural beauty into a Canyonlands waste dump? In response, DOE issued a report on site characterization and the national park which emphasized the local nature of the impacts. The national park was defined in terms of local recreation opportunities, physical features, and economic benefits—rather than a land use of national importance.[30] The significance of site characterization impacts was reduced by denigrating the uniqueness and significance of the national park itself:

> The ecological resources of the park are not, in general, unusual.
> . . . The archaeological resources of Canyonlands are not unusual

for those of the Colorado Plateau or for those of other area NPS [National Park Service] units.

. . . .

It is clear that the legislative mandates for many of these parks [national parks on the Colorado Plateau] are very similar to one another and to that of Canyonlands National Park. . . . Thus, the parks appear to have been established for similar general purposes.

. . . .

The positive impact that the repository and its personnel have on the local economy would offset those losses attributed to the loss in Canyonlands tourists.[31]

DOE estimated that site characterization impacts would affect only a few thousand visitors. Because these impacts would occur in a remote area, they could easily be mitigated by keeping people out of that part of the park. In any case, DOE concluded that the national parks of the region were similar and thus there were other, substitutable opportunities for the tourists. This ignored the fact that uniqueness, remoteness, and a pristine environment were three reasons why Congress declared the area a national park. Interpreting the impacts in local terms, DOE dismissed the national significance of site characterization activities and resisted demands that national meetings be held to air public concerns about constructing a nuclear waste dump adjacent to Canyonlands National Park. To concentrate on localized externalities within a host state, and to view them as the major impediment to siting, meant ignoring a broader, more diffuse range of impacts and source of public opposition.

By concentrating on the repository site, regional externalities could be excluded from consideration as well. For example, potential health and safety impacts along transportation corridors were treated generically in the environmental assessments. According to DOE, the "costs" of transporting waste five hundred versus one thousand miles were judged to be the same. Governor Lamm and State Senator Tom Glass of Colorado, commenting on the environmental assessments in a letter to the chairman of the NRC, challenged DOE's approach:

> As you know, the U.S. Dept of Energy (DOE) is now conducting Environmental Assessments at all of the potential repository sites. However, DOE is using a generic routing model in its assessments which fails to take into account critical factors affecting the specific routes which will be utilized. . . . We are very troubled that DOE would conduct in-depth studies in the environmental effects of disposing high-level waste at the various potential repositories but ignore the critical environmental effects of transporting the waste to each of those repositories. We are, of course, requesting DOE to expand its Environmental Assessments to include site-specific route information.[32]

Lamm and Glass went on to ask that the assessments include comparative analyses of the social and environmental costs associated with likely routes to each repository. They wanted an analysis of safety and other impacts produced by weather-related shipment delays which would concentrate spent fuel shipments in highway parking lots or railyards. In their opinion, DOE should address problems with inadequate state and local emergency response capabilities. Viewing externalities in strict, site-specific terms presupposed that some costs would be assumed by corridor states. For example, the responsibility for planning and developing the capability to respond to a radiological emergency was passed onto the states with much uncertainty about how it would be funded. By not tying sites to the regional transportation network, the environmental assessments included little geographically specific analysis upon which corridor states could comment. They were left with the more difficult task of trying to force DOE to recognize these potential externalities as relevant to site selection. Not until late 1986 did DOE begin to develop plans for transportation risk studies that would include state meteorological data, land-use patterns, and state-level accident data.[33] Thus, the regional extent of health and safety externalities associated with a specific site would not be known for years after a site was judged suitable for characterization.

The choice of sites also could be used to alter the definition of an unacceptable impact. In other words, different
locations captured various sensitivities to externalities. For
example, introducing a quantity of radioactive material at
a site in Wisconsin may represent a significant increase
above ambient levels of radiation. The same quantity introduced into the already polluted environment of Hanford,
Washington, or the Yucca Mountain, Nevada, site could represent a much smaller, perhaps insignificant increase which
may even be undetectable from the elevated background radiation levels at an already polluted site. In the past, populations living near existing nuclear reservations were likely
to be more tolerant of small, accidental releases of radiation;
whereas, at another location, the first time such a release
occurred it could prompt massive protests and demands that
the facility be closed. Thus, a legacy of externalities already
evident in the environments of Hanford and Yucca Mountain lowered political opposition to additional increments
of environmental degradation. A 1979 report of the comptroller general viewed these places as national sacrifice areas
suitable for waste disposal:

> Before the Department of Energy . . . selects any other repository
> site, it should give first consideration to determining if any of
> the existing, highly contaminated reservations are acceptable
> because
> • using them would avoid contaminating any more areas of the
> United States with radioactivity;
> • disposal of the Department of Energy generated waste would
> be simplified;
> • the sites are already federally owned, are in remote locations,
> and are in some cases so badly contaminated that they can never
> be returned to unrestricted uses; and;
> • public and political acceptance at these locations is likely to
> be higher than in other parts of the country.[34]

Radioactive wastes already stored at these locations set a
powerful precedent that added to political pressures to select one of them as the place for a repository. The comptroller general observed that, if DOE facilities such as Hanford

and the Nevada Test Site were found unsuitable for a repository, the Department of Energy would face harsh criticism
about the safety of radioactive waste already stored there.[35]
The continued operation of the vast nuclear complexes at
these sites could be threatened and new forms of regulatory
oversight demanded.

A repository location implied redistribution of a negative
externality and preservation of the benefits of nuclear
power. So long as spent fuel was stored at reactor sites—
concentrated in the eastern United States and often located
near urban areas—large numbers of people were exposed to
high-level waste (fig. 6). Widespread support existed for the
removal of spent fuel from reactor sites—an event welcomed by utilities. A repository located in a remote western
state, outside a utility company's service area, would concentrate the externalities associated with long-term exposure to high-level waste, remove a perceived health threat,
and still keep the benefits of nuclear reactors within the
utility's service area. Moreover, resistance to concentrating
high-level waste was likely only in isolated locations. Opponents would have to make their appeals through national
institutions where they would encounter constituents from
hundreds of jurisdictions who supported the removal of
spent fuel from local reactor sites. Additionally, states with
inventories of military and other high-level waste—West
Valley, New York, Savannah River, South Carolina, and Morris, Illinois—also looked to the repository for relief. For the
nuclear establishment, a repository meant that uranium
mining, concentrating, and enrichment facilities would continue to be needed because permanent disposal served the
once-through commercial fuel cycle. A repository might
even help restore support for the use of nuclear power. With
powerful states such as New York depending upon a repository, Nevada's veto was hardly a comprehensive insurance
policy.

In sum, little evidence suggests that the Yucca Mountain, Nevada, site was the outcome of a systematic site selection process. Rather, it should be seen as the product of

historical precedent and political-economic expediency. Similar to the emphasis on technical explanations of the problem, locational definitions of the nuclear waste dilemma masked fundamental and far-reaching conflicts. The push to settle on a site for a repository confined political participation and reduced the scope of issues, thus enhancing the nuclear establishment's power and control over future waste management activities. Emphasis on the choice of repository locations not only focused debate on a narrow range of issues and constituencies, it foreclosed discussion of broader issues related to the creation of a credible and legitimate process for developing a waste management strategy.

CHAPTER 8

CONCLUSION

To an outsider, the issue of what to do with high-level radioactive waste introduces a morass of obscure jargon and abstruse questions. An almost measureless bulk of documents, data, and technical reports describes the technology of nuclear waste management. The politics of nuclear waste involves hundreds of issues, actors, agencies, institutions, and jurisdictions. The policies guiding nuclear waste management are a mixture of legislative mandate, administrative interpretation, and programmatic precedent. While everyone can appreciate that a complex, highly sophisticated engineering is required to safely store nuclear materials for thousands of years, few have appreciated the political requirements necessary to design and implement such a solution. While vast resources have been expended on developing complex and sophisticated technologies, the equally sophisticated political processes and institutions required to develop a credible and legitimate strategy for nuclear waste management have not been developed. The history of high-level radioactive waste management describes repeated failure to recognize the need for institutional reform and reconstruction. Nuclear waste policy is the legacy of an unwillingness to discard the authority and commitments of the nuclear establishment.

The previous chapters used a broad perspective on the politics of nuclear waste repository to sort through the bewildering array of technical questions, issues, policies and politics. The alternative approach proposed at the outset of this analysis offers a means to appreciate the context and complexity of what at first look appear to be narrow, locational issues. The arguments developed in past chapters suggest that what appears to be a locational conflict may actually involve long-standing tensions that go far beyond questions of local impacts or compensation. The case of a nuclear waste repository offers lessons invaluable when assessing the political and institutional requirements of any process used to address the siting of other, controversial facilities. Furthermore, the evidence presented here suggests that challenges to the legitimacy of the existing repository program will continue unabated. Criticism of the priorities, scientific credibility, and legitimacy of the nuclear establishment appears to be achieving a momentum which may result in drastic restructuring of the nuclear industry and new forms of external control and oversight.

An Alternative Perspective

The perspective adopted here on the nuclear waste issue put aside the jumble of technical-legal questions crowding current discussion about the repository program, and instead focused on a simple question: how did we get here? Why did nuclear waste disposal become so controversial in the first place? After all, DOE had a waste management program long before Congress created a nuclear waste policy act. Why were waste disposal technologies, which had been around for decades, so difficult to implement? What produced and maintained this conflict? To address these questions, political-economic, social, historical, and geographic factors were considered as they influenced the following: the appearance and definition of high-level radioactive waste as a problem; the definition of a feasible and desirable solution to the problem, in both locational and technical terms; the push to embrace that solution in national legislation; and opposition to the implementation of that legislation.

Past approaches that analyzed conflicts, especially those involving the location of potentially hazardous facilities, were inadequate for explaining the complexity of issues and sources of opposition involved in the siting of a nuclear waste repository. For example, opposition to unwanted facilities is commonly viewed as a case of the NIMBY—"not-in-my-backyard"—syndrome. However, people may also organize in an attempt to capture the perceived benefits of such a facility. Political opposition and organization may occur on a variety of political-geographic scales. If one places too much emphasis on the locational—or site-specific—aspects of facility siting, only the most obvious frictions may be addressed. In the case of the repository, sources of conflict went much deeper than questions of local impact and mitigation, which explains the difficulty in resolving such cases by negotiation and compensation. A much wider set of factors must be considered when analyzing such confrontations. After tracing the origins of the nuclear power industry and the parallel evolution of nuclear waste as a problem, conflict over a repository site was found to be a continuation of long-standing tensions—between the priorities of a nuclear economy and health-environmental concerns; between national, bureaucratic control over nuclear policy and outsiders' attempts to affect those policies and regulate nuclear industries; and between the science used to defend the actions of the nuclear establishment and those of its opponents. Analysis of the repository situation points to the importance of considering precedent-setting interactions, especially as they figure in subsequent perceptions about the legitimacy, credibility, and priorities, of parties to a conflict. By focusing on the mechanics of site selection, analysts have only tinkered with superficial reforms. As much as anything, overlooking the issues of social priority, credibility, and legitimacy in assessments of what was wrong with the waste management program predisposed it to repeated failure.

Counter to the NIMBY argument, the evidence presented here sugests that opposition to an undesirable facility may not materialize at the local level where the negative impacts are assumed to be greatest. More often, opposition and support was geographically dispersed and fragmented. Contrary to some views, conflict involving the siting of hazardous facilities may be more than a case of the local versus the national state. Opposition, as well as support, may cut across local and national levels of the state, economy, and civil society. This geographical perspective is critical as it requires one to consider how the definition of a conflict as "locational" can circumscribe the topics viewed as suitable for political debate and the groups or individuals admitted to the political process. Attempts to confine this conflict to a limited set of locational issues ultimately were attempts to set the boundaries of political participation, interaction, and debate. More than a decision based on the geophysical merits of the site, the choice of Yucca Mountain, Nevada, for the repository was the geographic expression of political agendas, power, and well-entrenched political-economic relationships.

When looking at administrative programs, some policy analysts have suggested that conflict is due to incompatible views of programmatic details, or to irrationalities in the language of the law itself. But a historical analysis finds evidence of the conflict long before passage of the Nuclear Waste Policy Act. Indeed, as conflict is inherent in the process of making policy, it would be naive to assume that all differences are put to rest with the passage of a compromise piece of legislation. Rather, legislation may only change the forum and rules under which the same problems and issues will reappear.[1] In the case of nuclear waste, it was especially difficult to distinguish between policy formation and policy implementation given that the NWPA did not initiate a new repository program as much as endorse past plans and decisions. Concepts of power and agenda-setting point to ways in which members of the nuclear establishment affected policy makers' perceptions of legitimate issues and expertise, problem definitions and solutions. Analysts of modern bureaucracies and the politics of technology emphasize the importance of considering how reliance on expertise comes to define the language of political discourse. For these reasons, this analysis could not use passage of the NWPA as the starting point while remaining ignorant of the precedents and biases which drove its formation and subsequent implementation.

The contextual approach to nuclear waste politics permitted one to step back from the issues and events that preoccupy day-to-day politics and to recognize the context and origins of that politics. The gossip of the political animal who delights in the nuances of political maneuver, or the fine points of technical evidence and legal argument, offers little insight into the sources or resolution of longstanding, fundamental political problems. Despite the appeal of clean-and-easy explanations—especially those with acronyms—no single academic discipline or theory has a stranglehold on insight or explanation. Evaluation of policy failures and successes will be enriched by applying the multiple approaches and levels of analysis described in figures 3 and 4. The challenge is to recognize what each may offer and to use them to break out of old interpretations which perpetuate futile political discussions and processes and ineffective institutions.

A broader contextual orientation was necessary to address why conflict over the siting of a nuclear waste repository has persisted despite national legislation and repeated attempts to design and implement a repository program. The results of this analysis suggest that the conflict is likely to continue. National legislation and policies have repeatedly reasserted the priorities and authority of the nuclear establishment. Past policies directing management of high-level nuclear waste have left long-standing relationships among DOE, its contractors, and nuclear industries largely intact. Basic questions—about the wisdom of continuing to generate nuclear power and nuclear waste and the equity and injustice of using the federal mace to force a state that generates no nuclear power to accept nuclear wastes produced by private industry—were left unaddressed. Also untouched were basic questions about the continued credibility and legitimacy of the nuclear establishment. The historical evidence presented here suggests that, in the long term, the price of reassertion will be failure. The United States may never successfully site a repository—at least not until long-standing relationships among members of the nuclear establishment are dismantled and the political economy of nuclear power undergoes a massive restructuring. At the worst, the federal bludgeon—like the one already used in the site selection process—will continue to be used to raze criticism and opposition. But the questions will persist. In the end, the country may have a waste disposal facility which solves the utilities' problem; however, the public will have little confidence that such a facility is demonstrably safe or a technically defensible solution. The difficulty of undoing the massive mistake of siting a defective repository will eclipse even the enormous problems we now confront in disposing of spent fuel and other types of nuclear waste.

The Current Repository Program

The failure of the Nuclear Waste Policy Act and the Department of Energy's repository program was obvious as Congress debated the 1987 amendments to the NWPA. Nevertheless, the amendments, like their predecessor, confirmed the staying power of the nuclear establishment. When Congress revisited the NWPA to unlock yet another impasse in the repository program, it again resorted to incremental solutions which deferred to the nuclear establishment.

Any semblance of regional equity in the original site selection process was scrapped in what became known as the "Screw Nevada Bill." For years, various utilities had called the characterization of three sites too expensive and a waste of time. They got their wish. Work at other sites and consideration of a second repository in the eastern United States were halted. Even before completion of the characterization studies—which were supposed to establish the comparative, scientific basis for a site's suitability—Congress virtually mandated Yucca Mountain, Nevada, as the place for the repository. This eliminated comparisons of the relative effectiveness of different geologic formations and sites to isolate radioactive waste. The amendments made challenges to the site selection process and lawsuits over the environmental assessments moot. Like its predecessor, the amendments made few changes in the organization of institutions responsible for high-level waste management. Congress dismissed calls by Representative Morris Udall for a moratorium on waste disposal activities until the entire DOE program could be reevaluated. DOE remained the lead agency. The amendments called for a Nuclear Waste Technical Review Panel; but, two years later, the panel still lacked a full quotient of members, mainly because most of the candidates offered could not pass rules protecting against possible conflicts of interest.[2] In any case, the panel had no authority to halt the program or order actions on the part of DOE.

By prohibiting characterization studies at all sites except Yucca Mountain, Nevada, Congress selected an area adjacent to the DOE-controlled Nevada Test Site which many considered to be a politically acceptable waste disposal site. Lacking substantial information about geohydrologic and other conditions at the site, pundits continued the simplistic arguments of the past: Nevada's dryness and remoteness made it the ideal repository site; Nevada should accept the waste dump as needed diversification for an economy built on gambling. Unfortunately, such shabby arguments received a ready audience from powerful senators such as J. Bennett Johnston of Louisiana, who was determined to protect his state from a waste dump, serve the interests of the utility industry, and force the dump onto Nevada. The amendments offered a bribe: if the state forfeited its right to protest under the NWPA, it would receive ten million dollars a year before the repository opened and twenty million dollars a year thereafter. The state was also supposed to receive special consideration for federal projects such as the multi-billion-dollar Superconducting Super-Collider—but that project went to Texas anyway. Some speculated that a recession would make the repository look more attractive and predicted that the state legislature would be "willing to cut a deal" within two years.[3] (Just the opposite occurred. In 1989, the state legislature passed a law forbidding the storage or disposal of high-level waste in the state.)

At the same time, the amendments placed limitations on Nevada's ability to build alliances with states encompassing transportation corridors that would serve the repository. Section 175 of the amendments required DOE to report to Congress on "the potential impacts of locating a repository at the Yucca Mountain site," including environmental impacts produced by transportation activities. Despite requests from western governors, transportation, safety, and other issues were not considered at the regional level. The Department of Energy continued its practice of narrowly interpreting their charge and considered only site-specific impacts in the report.

Attempts to fragment opposition continued. DOE granted Clark County, Nevada, the status of an impacted county which meant that it could deal directly with DOE and receive money from the Nuclear Waste Fund. This meant the county no longer had to receive funding through the state nuclear waste office that openly opposed the repository. In 1989 the Nevada state legislature unequivocally stated its opposition to a repository by passing legislation which declared it "unlawful for any person or governmental entity to store high-level radioactive waste in Nevada." The constitutional questions raised by such a law will give Nevada a platform from which to argue that attempts to force a state to accept a repository for the disposal of commercial spent fuel—which serves no national purpose but mainly serves the purposes of profit-making corporations—is an abuse of federal power.[4] In the meantime, Senator Johnston, chairman of the Senate Energy Committee, inserted a provision in the fiscal year 1990 appropriations bill which cut, by sixty percent, Nevada's program for reviewing and verifying technical studies related to the repository project— virtually the only funds provided for state reviews mandated by the Nuclear Waste Policy Act. Full funding would be restored only after the secretary of energy certified that Nevada was cooperating with DOE.

In short, Nevada has been made the butt of a failed nuclear waste policy. The choice of Yucca Mountain as the site for a high-level waste repository had little to do with the ability of that site to isolate radioactive materials from the biosphere. Selection of that site said more about past investments by DOE and its contractors—and the influence of some powerful politicians—than it did about the site's technical superiority. Under the present circumstances, Yucca Mountain, Nevada, can never be considered the best, the safest, or the most technically sound site for nuclear waste disposal. The comparative analyses and site screening needed to make such a claim have never been performed. Rather, Congress' choice of Yucca Mountain as the site for a high-level radioactive waste dump continued the pattern of political and economic expediency at the expense of public health and safety. The 1987 amendments continued the pattern of deferring to the sites, technologies, expertise, and administrative programs favored by members of the nuclear establishment. Once again, in the push to answer the demands of nuclear utilities and industries, and reduce points of opposition to nonissues, politicians latched onto available sites and technology. While they did reject DOE's choice of Tennessee as the site for an MRS facility, in the case of the repository the amendments left past DOE decisions unchallenged. In the end, these policies have made substantial innovation—either technical or institutional— impossible. Work on other waste technologies and geologic formations suitable for a repository has been terminated, locking the country into only one waste disposal solution.[5]

State and local demands for changes in the waste management program have been largely unheeded. The amendments to the NWPA continued the pattern of federal bullying and reassertion of the nuclear establishment. They failed to address the nuclear establishment's loss of scientific credibility and legitimacy and were blind to the fact that public confidence in the nuclear establishment's ability to protect public health and the environment was lost. The crises and disorder which produced demands for congressional intervention in 1982, and again in 1987, have persisted.

The Future of the Repository Program and the Nuclear Establishment

The key question for the future of the repository program and high-level waste management is whether any program controlled by DOE can be viewed as scientifically credible or legitimate. The Department of Energy cannot escape its historical role as the promoter of nuclear energy, the tester of nuclear weapons, the agency whose contractors continued to fabricate nuclear weapons in an unsafe manner and ignored environmental, health, and safety regulations at government-owned facilities.

Within months after passage of the amendments to the NWPA, additional sources of information that compromised the integrity of the agency and its waste disposal program were being revealed. The General Accounting Office released a report which found that DOE's policy of suppressing negative information on potential repository sites was so pervasive that "DOE might never have publicly released the information if the Nuclear Regulatory Commission, through its on-site representative, had not identified and pursued the issue."[6] At Hanford, the GAO found that contractors operating DOE's nuclear waste tanks had failed to provide workers with adequate protection against exposure to asbestos.[7]

A report by a DOE geologist that raised serious questions about recent faulting, seismic, and volcanic activity at Yucca Mountain was not released until after the amendments to the NWPA were passed. DOE argued that the report was not released because it was going through an internal peer review process at the time Congress was targeting the Nevada site. However, the author of the report claimed that his questions about geologic problems, which could make the site unlicensable, had been forwarded to DOE management more than a year before. Another 1987 report, which was ignored while Congress homed in on the Yucca Mountain site, found that the probability of volcanic activity at the site was one hundred times greater than DOE estimates.[8] Other reports raised questions about DOE plans to place shaft openings in a floodplain.

The ability of DOE to manage a credible study of the Nevada site came under fire within a year after passage of the amendments. An investigation by the General Accounting Office found that quality control over data collection activities was so bad that they could derail the entire process of site characterization and licensing. The Department of Energy pointed to quality assurance plans on the shelf, but GAO said that little had been done to assure the proper implementation of such plans. The GAO was not alone in its criticisms. Increasingly, aggressive oversight of the DOE program by the Nuclear Regulatory Commission also raised questions about the organizational effectiveness of DOE and its ability to carry out quality work at the site (even as Congress was crippling Nevada's ability to monitor DOE studies). A team of hydrologists from the U.S. Geological Survey (USGS) charged that technical problems at Yucca Mountain could make the choice of that site "scientifically indefensible." Lack of autonomy by USGS scientists and their inability to conduct on-site work were cited as hindering the evaluation of DOE's work. Scientists at the Nuclear Regulatory Commission, the National Research Council and geologists from several universities, including Harvard, criticized DOE's analysis of groundwater problems as simplistic. The NRC's Advisory Committee on Nuclear Waste reproached DOE for neglecting studies which could detect potentially disqualifying features of the site.[9]

In the broader scheme, most damaging to the credibility

and legitimacy of DOE and its contractors, were revelations of widespread environmental problems at weapons facilities and attempts to cover up those problems.[10] Headlines screamed: "They Lied to Us" (*Time*, 31 October 1988); "Operators Got Millions in Bonuses Despite Hazards at Atom Plants" (*New York Times*, 26 October 1988); "U.S. for Decades Let Uranium Leak at Weapons Plant" (*New York Times*, 15 October 1988); "Cleanup of Nuke Plants Put at $130 Billion" (*Rocky Mountain News*, 14 July 1988); "Groundwater Tainted Near the Nation's 16 Nuclear Weapons Plants, Data Shows" (*New York Times*, 7 December 1988); "Rocky Flats Illegally Burned, Dumped Waste, U.S. Claims" (*Denver Post*, 10 June 1989). In 1989, the Federal Bureau of Investigation raided DOE's Rocky Flats Nuclear Weapons Plant as part of a criminal investigation into the cover-up of environmental violations at the plant. The crisis in DOE's weapons and research facilities had reached the point where newly appointed Secretary of Energy James Watkins called for a new culture of accountability and blamed past problems on an ingrained DOE culture which viewed production goals as incompatible with public health and environmental protection. In his words,

> So, now, the chickens have finally come home to roost and years of inattention to changing standards of and demands regarding the environment, safety and health are vividly exposed to public examination, almost daily. I am certainly not proud or pleased with what I have seen over my first few months in office. . . .
>
> Since undertaking my present assignment as Secretary of Energy only four months ago, I have also been surprised to learn that the Department relies on insufficient scientific information in making its decisions and in developing public policy. In this regard, I am instituting measures that will greatly increase the roles State agencies, the Environmental Protection Agency . . . play in DOE decision-making to provide a greater influence on the quality of the scientific data we employ to make our decisions affecting public health, safety, and the environment.[11]

Watkins as much as accused DOE management of deceiving him, saying, "Because of the serious nature of the many management problems facing me at DOE, I have found that I must undertake my own assessment of all DOE operations in order to come up with an adequate baseline of information, one upon which I can then make informed judgments."[12] Watkins called for independent validation of environmental data. He terminated DOE's "non-uniform, haphazard, overly-decentralized, and self defeating" approach to implementing—more accurately, avoiding implementation of—the National Environmental Policy Act. He cited his own agency for a lack of "sufficient numbers of appropriately skilled DOE line supervisors." He called for a Comprehensive Epidemiologic Data Repository in which past and present information on DOE workers would be placed and made accessible to outside researchers, reversing the practice of refusing to give researchers unaffiliated with DOE access to such data. He called for inspections of DOE facilities by the Occupational Safety and Health Administration and dispatched his own teams to inspect and collect information on environmental problems at thirty-five DOE facilities.[13]

Problems and delays in opening the Waste Isolation Pilot Plant in New Mexico further tested confidence in the agency's ability to mange a waste disposal program. Watkins referred to the WIPP program as "a classic example of the crying need to re-establish a well-aired and documented baseline of understanding."[14] He said that the agency could no longer afford to continue its "blind allegiance to past decisions" at the expense of environmental protection.

If serious problems permeated the weapons program and the program for opening a military waste repository, couldn't these same problems exist in DOE's civilian programs? The secretary's criticisms indicated that environmental problems in the weapons program were an outcome of the culture of the nuclear establishment, the organizational structure of DOE, and complacent relationships between DOE and its contractors—factors which could also apply to the repository program. How could the public and the state of Nevada be assured that these same problems did not infest the repository program? If DOE's weapons program merits such close scrutiny, perhaps the civilian program should not be slighted. Missed deadlines, delays in the program, lack of leadership, charges of conflict of interest in the award of contracts, all point to significant management problems within the Office of Civilian Radioactive Waste Management. There is no shortage of problems that could furnish ammunition for the same sort of criticism being leveled at the weapons program.

Mounting problems in the nuclear establishment will make continued reassertion of the policies and priorities of the establishment increasingly difficult. The merits of past waste management decisions and commitments may no longer be assumed. DOE is likely to have a difficult time maintaining the authority and interests of members of the nuclear establishment as support grows for new sources of authority and forms of oversight. A flood of legislation calling for independent oversight of DOE was proposed in the One-hundred-first Congress. The future may see members of the nuclear establishment increasingly regulated by the EPA, independent commissions, state health departments, and other agencies.

Growing rifts within the nuclear establishment itself may change the future landscape of nuclear policy making. The relationships and networks that have persisted since World War II may prove less and less workable. For example, nuclear utilities are purchasing less reactor fuel from DOE. This comes at a time when some members of Congress are trying to force utilities to pay the full cost of DOE-provided uranium enrichment services, including the cost of decommissioning old plants. Despite claims by the nuclear industry-sponsored U.S. Council for Energy Awareness that nuclear power is the key to the U.S. energy security, utilities dramatically increased their purchase of enriched uranium from the Soviet Union between 1986 and 1988.[15] In an unprecedented move, several utilities also announced plans to build their own $750 million enrichment facility in the home state of Senator J. Bennett Johnston.

Criticism by utilities fed up with delays in the repository program have grown especially harsh. Commenting on continued delays despite the amendments to the NWPA intended to accelerate the program, the director of the Utility Nuclear Waste Management Group, said: "I think the best thing to say is that the utilities are pretty fed up with DOE. . . . They are fed up with the lack of progress. . . . Nothing has changed and I think that's what really infuriated the utilities."[16] He criticized DOE for being unresponsive to industry's offers of suggestions and assistance on quality assurance and other issues. He cited the failure of DOE to maintain leadership and said the nuclear industry has given up on DOE's promise of a repository by 2003 (five years later than the originally scheduled 1998 opening). More and more utilities are now planning their own interim solution to the storage problem by developing new reactor-site storage facilities for spent fuel rods.

This development raises the question of the near-term need for a repository. Perception of the need for rapid deployment of a repository has driven the high-level waste program since the days a repository was proposed for Lyons, Kansas. That perception has been used to justify eliminating site inventories, limiting in-depth studies, and steamrolling state concerns. It is time to put this argument to rest. The Nuclear Regulatory Commission recently concluded that utilities possess the capacity to store all spent fuel at reactor sites until the year 2025. Only then would a repository be needed. The implications are clear. DOE does not need to accept spent fuel by 1998. The United States does not need to force the construction of a repository by 2003. In short, there is time to reexamine the premises of the NWPA and to assess the failure of current policy and institutions.

In the future, the Nuclear Regulatory Commission may prove to be a more potent, independent source of criticism of DOE and its contractors. NRC has shown indications of becoming increasingly critical of DOE plans and is intensifying its oversight of DOE's site characterization program. In the end, the NRC must authorize construction of the repository and issue the operating license for it. Long battered by environmental and citizen groups for taking a complacent attitude toward nuclear power, the commission may assert its independence and distance itself from an unpopular and unworkable repository program. The NRC may emerge as the wild card of opposition. After all, the commission has every incentive to take a hard, critical look at the DOE program to make sure that no one blames the NRC for opening a flawed repository.

Maintaining that its contractors are shielded from environmental regulations, DOE has defended them in cases before the EPA and the states: it has reimbursed them for fines paid and legal costs associated with contesting environmental regulations. However, more and more members of Congress are convinced that it's time to pull contractors out from behind the protective skirts of DOE. Various attempts have been made to extend EPA's and states' authority over DOE and its contractors by making them subject to the same laws and regulations that apply to commercial facilities storing or disposing of hazardous wastes. DOE has argued that it already requires its facilities to meet safety standards comparable to commercial facilities. However, a GAO investigation found DOE standards to be unclear and incomplete, and that DOE lacked a formal, systematic program for assessing compliance.[17] Despite some evidence of change, key relationships among members of the nuclear establishment remain intact. In a survey of thirteen key votes by Congress, the Washington, D.C.-based Public Citizen (a Ralph Nader activist group) found that the nuclear industry's clout in Congress had actually increased.[18] Many members of Congress have continued to promote nuclear power by backing legislation which would streamline licensing commercial reactors and funding the development of a new generation of reactors. Thus, it would be premature to view the country as being on the verge of a comprehensive revamping of national policies and programs which have favored the development of nuclear power. New institutions, patterns of participation, and sources of power are not imminent. But perhaps, in the face of the disaster at DOE, calls for new social priorities, new institutions, and broader participation in the nuclear waste program will at least be viewed as legitimate concerns and issues.

Bungled attempts to site a repository will not soon be forgotten. DOE's inability to produce a credible or legitimate site selection process has meant that its program could only be defended by an act of Congress. In the meantime, a broad range of U.S. citizens became sensitized to the problem of toxic and radioactive waste disposal and the ineptness of DOE to address those problems. It will now be even more difficult to convince the public of the value of any technology, nuclear or otherwise, that prompts such conflict over the disposal of its by-products. Given the public distrust raised by DOE actions, the nuclear establishment will likely bear even greater scrutiny of its activities in the future. Calls for reform and reconstruction will not disappear. The lessons to be learned from DOE's experience offer a warning to others seeking solutions to complex problems involving the disposal of toxic materials or the siting of hazardous industrial facilities.

Implications for the Siting of Any Controversial Facility

The conflict over siting a nuclear waste repository offers several lessons relevant to future attempts to address controversies over locational and technological issues. At the least, efforts to foresee or resolve these types of conflicts must consider that the following issues may be present:

Credibility issues. The conflict may involve access to, control of, and confidence in the expertise used to establish the scientific and technological credibility of proposed solutions.

Legitimacy issues. The conflict may involve demands for expanded access to decision-makers and broader participation in the political process. The equity or fairness of the process used to define a problem and identify a solution may also be contested—long before the outcome is known. Participants may view state authority as serving the public interest, or as merely serving a special interest group and preempting local concerns.

Economic and social priorities. The conflict may involve differing perceptions of the severity of the problem, the need for federal intervention, the timeliness of the settlement, and the parties responsible for financing the solution. Dissenting priorities of environment and public health, and those of the economy and the national security, may be inherent in the parties' political positions.

These crises may produce calls for new institutions, inventive forms of political participation and decision making, public access to technical expertise, and other changes which could alter the balance of power and relationships among existing groups and institutions. Solutions which do not address these issues and which only serve to entrench existing processes and powers, are predisposed to failure. At the least, if unresolved, these crises will set the stage for continuation of the conflict in other forums. One may win the battle on technical or legal grounds but in doing so may irretrievably lose public confidence and support.

Federal preemption and overriding state and local opposition is a solution which holds the allure of a quick and final end to a conflict; in practice, it often fails to achieve the hoped-for finality. This study suggests that attempts to resolve locational and technological conflicts should confront the questions of legitimacy, credibility, and economic priorities head-on. Rather than debating the merits of a specific technical-locational solution which must then be forced on a local population, one should focus on creating a legitimate and credible process for defining the problem and its solution. Questions of technical uncertainty, for example, should be addressed by independent technical review bodies composed of members from a wide range of disciplines, citizen groups, and outlooks. "Independent" means that the data and studies used to support their analysis should not be controlled by a group with an interest in promoting a particular result. Scientific credibility will be compromised if it is collected and interpreted by an agency interested in promoting a technology or location. An independent review group should have the authority to stop any development until the issues are resolved.

The threat of economic crisis may push policy makers to trade "safety checks," such as environmental impact statements, for the chance at a speedy solution; in doing so, they may neglect to consider the costs and delays produced by the bungled solution. Predictions of financial disaster in the utility industry were a powerful, immediate force driving the NWPA. However, years later, and still without a repository, that disaster has not materialized. A sound approach would give more consideration to the long-term interests of public assurance and environmental quality than it would to the short-term interests of an industry.

A legitimate decision-making process will accommodate demands for participation by groups committed to the issue. Exclusion of environmental groups such as the Natural Resources Defense Council and transportation corridor states from the repository program was inexcusable. Any attempt to address the notion of legitimacy must consider the legacy of public trust (or distrust) in existing institutions. It must recognize the drawbacks of relying on a mission-oriented government agency and its clientele group to outline a new decision-making process. Particularly controversial issues—especially those which existing agencies have failed to resolve—may require completely new institutions. Such institutions may be intergovernmental, interstate, or inter-agency—a mixture of the civil, economic, and governmental. They may cut across common hierarchies and notions of government and public administration, and their form may vary with context—the history, geography, politics, and economics of the issue.

The publicly acceptable resolution can not be discovered and certainly not imposed by technical experts. No stock set of qualities or components can be used to construct such a solution in the laboratory. The legitimate, valid, and equitable solution can not be defined apart from the process used to create it. Thus, the challenge and responsibility for policy makers is to initiate a widely supported political process capable of producing programs, management structures, and solutions in which all parties have a stake in their success.

NOTES

BIBLIOGRAPHY

INDEX

NOTES

INTRODUCTION

1. This story is recounted in G. Clarfield and W. Wiecek, *Nuclear America: Military and Civilian Nuclear Power in the United States, 1940–1980* (New York: Harper & Row, 1984), pp. 349–50.

2. The politics of civilian high-level waste is itself a huge topic without attempting to cover the related topics of military high-level and transuranic wastes or commercial low-level wastes. As used in this study, high-level waste refers only to spent fuel rods and other highly radioactive materials produced by fission in commercial nuclear reactors. Nuclear reactors, as well as medical and scientific sources, produce millions of cubic meters of low-level wastes. They are not as radioactive, decay much more quickly, and are currently disposed of in surface facilities. One of the most serious problems of military waste disposal involves treating the liquid by-products of spent fuel reprocessing. Once this waste is stabilized (vitrified), the commercial repository will also be used for this military high-level waste, but it will constitute a relatively small percentage of the total waste inventory—probably less than twenty percent. In any case, the military did not drive the development of a high-level waste repository, preferring instead to reprocess spent reactor fuel, recover the plutonium for use in weapons programs, and store wastes on site.

Compared with commercial high-level waste, policies governing military waste and low-level waste disposal have different origins and histories. They emphasized different sets of political-economic factors, institutions, and political groups. The disposal of military waste offered much less opportunity for state oversight or involvement. On the other hand, policy on low-level waste disposal evolved toward a state-oriented, regional approach. There are, of course, crossover issues—questions of institutional reform, legitimacy, scientific credibility, local control versus regional ap-

proaches, national preemption of local authority. However, addressing them is beyond the scope of this study. And while it would be interesting to contrast the development of these different policy tracks, such an effort is for another study.
3. *Nucleonics Weekly*, 20 May 1976. Reprinted in House Subcommittee on Energy and the Environment, *Hearings: Nuclear Waste Disposal in Michigan*, 94th Cong., 2d sess., 1976, pp. 267–68.
4. In December 1987, Congress amended the 1982 Nuclear Waste Policy Act and selected Yucca Mountain, Nevada, as the only site that could be studied as suitable for a first repository. Site characterization studies at Hanford, Washington, and Deaf Smith, Texas, were terminated. Investigations of sites for a second repository in the eastern United States were also prohibited. While this study focuses on the 1982 act and the attempts to implement it, the 1987 amendments had the effect of further reinforcing the priorities, problems, and technical solutions laid out in the 1982 act.
5. For a discussion of these issues, see L. Carter, *Nuclear Imperatives and Public Trust: Dealing with Radioactive Waste* (Washington, D.C., 1987).

CHAPTER 1. *A Perspective on the Politics of Nuclear Waste*

1. G. Clarfield and W. Wiecek, *Nuclear America: Military and Civilian Nuclear Power in the United States, 1940–1980* (New York: Harper & Row, 1984), pp. 175–76.
2. The term *establishment* is suggested by at least three works on the history of nuclear power: Metzger's condemnation of the AEC, *The Atomic Establishment* (New York: Simon & Schuster, 1972); Bertrand Goldschmidt, *The Atomic Complex* (La Grange, Ill.: American Nuclear Society, 1982); and Richard Lewis's work on citizens versus the atomic-industrial establishment, *The Nuclear Power Rebellion* (New York: Viking, 1972). Some have also referred to the nuclear-industrial complex as a subgovernment. See J. Temples, "The Politics of Nuclear Power: A Subgovernment in Transition," *Political Science Quarterly* 95 (1980): 239.
3. A. Kirby, "State, Local State, Context and Spatiality," Institute of Behavioral Science, Boulder, Colo., 1987. Photocopy.
4. R. Alford and R. Friedland, *Powers of Theory: Capitalism,*

the State and Democracy (Cambridge: Cambridge University Press, 1985), p. 3.

5. Alford and Friedland, *Powers of Theory*, p. 7; see also S. Lukes, *Power: A Radical View* (London: Macmillan, 1974).

6. D. Torgerson, "Contextual Orientation in Policy Analysis: The Contribution of Harold D. Lasswell," *Policy Sciences* 18 (1985): 241–61.

7. W. Lanouette, "Atomic Energy 1945–85," *Wilson Quarterly* 9, no. 5 (1985): 91–131.

8. M. Dear and G. Clark, "The State and Geographic Process: A Critical Review," *Environment and Planning—A* 10 (1978): 182.

9. See, for example, M. Dear, "A Theory of the Local State," in *Political Studies from Spatial Perspectives*, ed. A. Burnett and P. Taylor (Chichester: J. Wiley, 1981), pp. 183–200.

10. M. Boddy, "Central-Local Government Relations: Theory and Practice," *Political Geography Quarterly* 2, no. 2 (1983): 134–36.

11. See chapter 7, and the discussion later in this chapter on local support for nuclear industries, including the state of New York's sponsorship of a spent fuel reprocessing facility, a rural Utah town's solicitation of a repository, and strong local support for nuclear industries in eastern Washington's Tri-Cities.

12. R. Arnold, *Congress and the Bureaucracy: A Theory of Influence* (New Haven: Yale University Press, 1979).

13. P. Bachrach and M. Baratz, "Two Faces of Power," *American Political Science Review* 56 (1962): 948.

14. DOE's discretion to define key technological and locational issues is a major source of conflict and opposition which will be discussed in chapters 6 and 7. Such discretion has also been an issue for "expert" federal agencies involved in forest, range, irrigation, hydroelectric power, and other aspects of resource management.

15. G. Benveniste, *The Politics of Expertise* (Berkeley: Glendessary Press, 1972).

16. J. Gaventa, *Power and Powerlessness: Quiescence and Rebellion in an Appalachian Valley* (Urbana: University of Illinois Press, 1980), p. 15.

17. P. Slovic and G. Fischhoff, "How Safe is Safe Enough?" in *Too Hot to Handle?: Social and Policy Issues in the Management of Radioactive Waste*, ed. C. Walker, L. Gould, and E. Woodhouse (New Haven: Yale University Press, 1983), pp. 112–50.

18. Gaventa, *Power and Powerlessness*, p. 255.

19. S. Hadden, J. Chiles, P. Anaejionu, and K. Cerny, *High Level Nuclear Waste Disposal: Information Exchange and Conflict Resolution* (Austin: Texas Energy and Natural Resources Advisory Council, 1981).

20. A representative example of this position is found in testimony by Sherwood Smith, Chief Executive Officer, Carolina Power and Light, who represented the American Nuclear Energy Council, Edison Electric Institute, and the Utility Waste Management Group in hearings before Congress. House Subcommittee on Energy and Power, *Hearings: Nuclear Waste Disposal*, 96th Cong., 2d sess., 1980, 184, 189–91.

21. A. Kirby, *The Politics of Location* (London: Methuen, 1982), pp. 156–57.

22. This function of the state was evident in the growth of federal regulation of the U.S. economy that occurred in the first years of the twentieth century. See G. Kolko, *The Triumph of Conservatism* (New York: The Free Press, 1963), p. 6.

23. J. Pressman and A. Wildavsky, *Implementation*, 3rd ed. (Berkeley: University of California Press, 1984), pp. 133–35.

24. R. Rudolph and S. Ridley, *Power Struggle: The Hundred-Year War Over Electricity* (New York: Harper & Row, 1986), p. xi.

CHAPTER 2. *The Nuclear Establishment*

1. For works that tend to be less partisan, see the literature reviews and bibliographies included in: W. Lanouette, "Atomic Energy, 1945–1985," *The Wilson Quarterly* 9, no. 5 (1985): 132–33, which will also lead the reader to earlier literature reviews; S. Del Sesto, *Science, Politics, and Controversy: Civilian Nuclear Power in the United States, 1946–1974* (Boulder, Colo.: Westview Press, 1979); G. Clarfield and W. Wiecek, *Nuclear America: Military and Civilian Nuclear Power in the United States 1940–1980* (New York: Harper & Row, 1984), which includes abstracts of older works; Office of Technology Assessment, *Nuclear Power in an Age of Uncertainty* (Washington, D.C., 1984); N. Moss, *The Politics of Uranium* (New York: Universe Books, 1982); J. Campbell, *Collapse of an Industry: Nuclear Power and the Contradictions of U.S. Policy* (Ithaca, N.Y.: Cornell University Press, 1988).

2. Two journalists, D. Barlett and J. Steele, have written *Forevermore: Nuclear Waste in America* (New York: W. W. Norton, 1985), which attempts to comprehensively discuss nuclear waste disposal. Written for a popular audience, it leads the reader to many key federal documents. Other works to consult include: *Essays on Issues Relevant to the Regulation of Radioactive Waste Management*, ed. W. Bishop et al. (Washington, D.C., 1978); *The Politics of Nuclear Waste*, ed. E. W. Colglazier (New York: Pergamon Press, 1982); *Too Hot to Handle?: Social and Policy Issues in the Management of Radioactive Wastes*, ed. C. Walker, L. Gould, and E. Woodhouse (New Haven: Yale University Press, 1983); R. Lipschutz, *Radioactive Waste: Politics, Technology, and Risk* (Cambridge, Mass.: Ballinger, 1980), a report of the Union of Concerned Scientists; Office of Technology Assessment, *Managing Commercial High-Level Radioactive Waste* (Washington D.C., 1982); L. Carter, *Nuclear Imperatives and Public Trust: Dealing with Radioactive Waste*, (Washington, D.C., 1987).

3. Del Sesto, *Science, Politics, Controversy,* pp. 76–77. See also J. Morone and E. Woodhouse, *The Demise of Nuclear Energy?* (New Haven: Yale University Press, 1989), p. 63, for a list of federally subsidized reactor demonstration projects.

4. Clarfield and Wiecek, *Nuclear America,* p. 194.

5. Del Sesto, *Science, Politics, Controversy,* p. 122.

6. In the 1970s, electrical utilities became involved in mining coal, lignite, and even precious metals. In the 1980s, it has become more difficult to define a utility in terms of electrical generation as the industry continues to diversify and restructure itself. An increasing percentage of their profits now comes from real estate development and other nonelectrical production ventures. See S. Fenn, *America's Electric Utilities: Under Siege and in Transition* (New York: Praeger, 1984), pp. 95–105. "Their" power plants are often owned by consortiums of private investors and leased back to the utility. It is difficult to precisely define a nuclear utility since utilities may have a mix of nuclear and conventionally fueled plants.

7. Fenn, *America's Electric Utilities,* pp. 6–7.

8. The accident was the topic of John Fuller's popular account, *We Almost Lost Detroit* (New York: Reader's Digest Press, 1975).

9. House Subcommittee on Energy Conservation and Power, *Hearings: Nuclear Waste Disposal Policy* (97th Cong., 2d sess., 1982), pp. 430–31.

10. As one example, see Bartlett and Steele, *Forevermore,* pp. 30–35. Using a variety of federal government sources, the authors composed a list of radioactive waste storage and disposal sites. Although their list is not comprehensive—overlooking many mill tailings sites—it does suggest the geographical extent of the problem and the number of jurisdictions with an interest in a solution.

11. The National Academy of Sciences' Committee on Waste Management endorsed burial of high-level waste in salt as the most promising method in 1957 and again in 1961. These two reports set the conceptual foundations of future research on the waste disposal problem. Twenty years later, they were still being used to justify waste disposal decisions. See D. Metlay, History and Interpretation of Radioactive Waste Management in the U.S.," in *Essays on Issues Relevant to the Regulation of Radioactive Waste,* ed. W. Bishop et al., p. 4.

12. The AEC cut off its support for research on loss of coolant accidents—a likely source of a reactor meltdown. Various reports by Robert Gillette in *Science* in 1972 discussed conflict of interest and collusion between the AEC and industry in setting research priorities. In the 1970s, the scientist who revealed higher than usual incidents of cancer in Hanford workers found funds for his research cut off and his data turned over to AEC analysts. S. Hilgartner, R. Bell, and R. O'Connor, *Nukespeak: The Selling of Nuclear Technology in America* (New York: Penguin Books, 1983), pp. 104–08.

13. D. Metlay, "History and Interpretation of Radioactive Waste Management in the U.S.," in *Essays on Issues Relevant to the Regulation of Radioactive Waste,* ed. W. Bishop et al., pp. 1–19.

14. Johnson and Zeigler have mapped standard metropolitan statistical areas (SMSAs) located within fifty miles of one to two and three to eight nuclear reactors. Particularly vulnerable areas include SMSAs around the Great Lakes and in the Boston-Washington, D.C., corridor. J. Johnson and D. Zeigler, "Evacuation Planning for Technological Hazards," *Cities* 3, no. 2 (1986): 154. Remote siting would increase electricity transmission costs and reduce utilities profits. See J. Campbell, *Collapse of an Industry: Nuclear Power and the Contradictions of U.S. Policy* (Ithaca, N.Y.: Cornell University Press, 1988), p. 52.

15. House Subcommittee on Energy Conservation and Power, *Hearings* (1982), p. 221.

16. House Committee on Science and Technology, *Nuclear*

Waste Research, Development, and Demonstration Act of 1980, 96th Cong., 2d sess., 1980, H. Rept. 96–1156, pt. 1. Additional discussion and references on this point can be found in chapter 7.

17. For examples of interpretations of locational conflict, see K. Cox, "Residential Mobility, Neighborhood Activism and Neighborhood Problems," *Political Geography Quarterly* 2, no. 2 (1983): 99–118. Also see various works by Wolpert cited in the bibliography. Cases supporting this viewpoint can be found in Environmental Protection Agency, *Siting of Hazardous Waste Management Facilities and Public Opposition* (Washington, D.C., 1979); T. Gladwin, "Patterns of Environmental Conflict Over Industrial Facilities in the U.S.—1970–1980," *Natural Resources Journal* 20 (1980): 243–74.

18. F. Shelley and G. T. Murauskas, "Local Conflict, Local Autonomy, and Regional Concern: Siting Nuclear Waste in South Dakota," presented at the annual meeting of the Association of American Geographers, Detroit, Mich., April 1985. Also see G. T. Murauskas and F. Shelley, "Local Political Responses to Nuclear Waste Disposal," *Cities* 3, no. 2 (1986): 157–62.

19. Office of Technology Assessment, *Nuclear Power in an Age of Uncertainty* (Washington, D.C., 1984). See page 216 for a list of states with laws and regulations restricting construction of nuclear power plants.

CHAPTER 3. *Disorder and Crisis*

1. S. Macgill, "Exploring the Similarities of Different Risks," *Environment and Planning—B* 10 (1983): 303–29.

2. Cambridge reports cited in Office of Technology Assessment, *Nuclear Power in an Age of Uncertainty* (Washington, D.C., 1984), pp. 211–47.

3. F. Sandbach, *Environment, Ideology and Policy* (Oxford: Basil Blackwell, 1980), chs. 5 and 6.

4. A. Lovins, "Energy Strategy: The Road Not Taken," *Foreign Affairs* 55, no. 1 (1976): 65–96.

5. V. Covello, "The Perception of Technological Risks: A Literature Review," *Technological Forecasting and Social Change* 23 (1983): 285–97.

6. A similar theme is developed in S. Skowronek, *Building a*

New American State: The Expansion of National Administrative Capacities, 1877–1920 (Cambridge: Cambridge University Press, 1982).

7. Senate Committee on Governmental Affairs, *Hearings: Nuclear Waste Management Reorganization Act of 1979*, 96th Cong., 1st sess., 1980, pp. 35–37.

8. D. Ford, *The Cult of the Atom: The Secret Papers of the Atomic Energy Commission* (New York: Simon and Schuster, 1982). Portions of the book were published earlier in the *New Yorker*.

9. P. Huber, "The Bhopalization of American Tort Law," in *Hazards: Technology and Fairness*, ed. National Academy of Engineering (Washington D.C.: National Academy Press, 1986), pp. 89–110.

10. S. Carnes, "Confronting Complexity and Uncertainty: Implementation of Hazardous Waste Management Policy," in *Environmental Policy Implementation*, ed. D. Mann (Lexington, Mass.: Lexington Books, 1982) pp. 35–50.

11. For a list of state and local regulations governing nuclear transportation, most passed since the mid-1970s, see T. Overcast and B. Schuknecht, *Federal Preemption of State and Local Nuclear Transportation Regulations* (Albuquerque: Sandia National Laboratories Transportation Technology Center, 1985). For a list of the siting laws twenty-six states passed in the 1970s, see Office of Technology Assessment, *Nuclear Power*, p. 152.

12. Two very different, but often complementary, interpretations of problems in the nuclear utility industry are C. Flavin, *Nuclear Power: The Market Test* (Washington, D.C.: Worldwatch Paper #57, 1983); J. Cook, "Nuclear Follies," *Forbes*, 11 February 1985, pp. 82–100.

13. Senate Committee on Governmental Affairs, *Hearings*, 1980, p. 36.

14. B. Ackerman and W. Hassler, *Clean Coal/Dirty Air* (New Haven, Conn.: Yale University Press, 1982), pp. 1–12.

15. F. Rourke, *Bureaucracy, Politics, and Public Policy*, 3rd rev. ed. (Boston: Little, Brown, & Co., 1984), p. 15.

16. Ibid., p. 20.

17. G. Beneviste, *The Politics of Expertise* (Berkeley: Glendessary Press, 1972), p. 29.

18. D. Metlay, "Radioactive Waste Management Policymaking," in *Managing the Nation's Commercial High-level Ra-*

dioactive Waste, Office of Technology Assessment (Washington, D.C., 1985), p. 202. See also D. Metlay, "History and Interpretation of Radioactive Waste Management in the U.S.," in *Essays on Issues Relevant to the Regulation of Radioactive Waste Management,* ed. W. Bishop et al. (Washington, D.C., 1978), p. 4.

19. S. Nealey and J. Herbert, "Public Attitudes Toward Radioactive Wastes," in *Too Hot to Handle?: Social and Policy Issues in the Management of Radioactive Wastes,* ed. C. Walker et al. (New Haven: Yale University Press, 1983), pp. 95–96.

20. Testimony of a spokesman for Edison Electric Institute: House Subcommittee on Energy and the Environment, *Hearings: To Amend the Atomic Energy Act of 1954,* 96th Cong., 2d sess., 1980, 644–45. Also see testimony of Boston Edison Co. and the utilities' Nuclear Waste Management Group: House Subcommittee on Energy Research and Production, *Hearings: H.R. 7418—Nuclear Waste Research Development and Demonstration Act of 1980,* 96th Cong., 2d sess., 1980, 98–110.

21. Testimony of a spokesman from General Electric for the Atomic Industrial Forum: House Subcommittee on Energy and the Environment, *Hearings: Radioactive Waste Legislation,* 97th Cong., 1st sess., 1981, 572–75.

22. House Subcommittee on Energy and the Environment, *Hearings: Nuclear Waste Disposal in Michigan,* 94th Cong., 2d sess., 1976.

23. *Congressional Quarterly,* 18 March 1978, p. 734.

24. House Subcommittee on Energy and the Environment, *Hearings: High-level Nuclear Waste Management,* 97th Cong., 2d sess., 1982, 35.

25. S. Hilgartner, R. Bell, and R. O'Connor, *Nukespeak: The Selling of Nuclear Technology in America* (New York: Penguin Books, 1983), p. 139.

26. S. Rothman and S. Lichter, "Elites in Conflict: Nuclear Energy, Ideology, and the Perception of Risk," *Journal of Contemporary Studies* 8, no. 3 (1985): 23–44.

27. For a description of these diverse targets of citizen activism, see chap. 6, "Citizen Rebellion," in R. Rudolph and S. Ridley. *Power Struggle: The Hundred-Year War Over Electricity* (New York: Harper & Row, 1986), pp. 144–79.

28. G. Downey, "Politics and Technology in Repository Siting: Military Versus Commercial Nuclear Wastes at WIPP, 1972–1985," *Technology in Society* 7 (1985): 47–75. The Waste Isolation Pilot

Plant offers insight into some of the problems which would later plague the commercial high-level waste repository. In a postscript, Downey briefly discusses how the 1982 NWPA was an attempt to resolve the problem of legitimizing siting decisions by establishing a federal/state decision-making procedure that would eventually require a national consensus on each site selection decision— expressed as Congress' willingness to either override or sustain a state's notice of disapproval (that is, the state veto). However, because WIPP was a military facility, it did not encompass the same type of legal, institutional, and procedural issues that developed with a civilian repository. For example, the WIPP site was chosen by the AEC in 1972 as part of an administrative waste management program. It was never part of a legislatively mandated site selection or evaluation process. The project did not attract the attention which was focused on the NWPA repository, though at times it was still controversial. Initially, New Mexico's governor agreed to support a preliminary study of the site, and a later governor refused to demand a state veto over the project. Local support for the project was high and inaction by the state legislature suggested tacit approval. The state did not acquire any formal authority to participate in decision making; thus, the legal basis for state opposition was severely limited. Military lobbyists and the powerful armed services committees within Congress overrode President Carter's attempt to halt the project as well as proposals to use WIPP as a spent fuel repository—which could have increased civilian and other congressional committees' oversight of the project. The military nature of the project was used to circumvent NRC and other procedural and environmental standards. After the election of President Reagan, construction on the project proceeded quickly. Finally, the populations benefiting from a military repository are very different from the beneficiaries of a commercial waste facility. The facility was set to open in 1988 but questions about underground waste disposal tests at the site, the safety of containers used to transport the transuranic waste, and approval by the Environmental Protection Agency of the site for the storage of hazardous chemical wastes—which have been mixed with radioactive wastes—have delayed its opening.

29. These views of legitimacy, containing elements of Weber and Habermas, were stimulated by J. Bensman, "Max Weber's Concept of Legitimacy," in *Conflict and Control: Challenge to Legitimacy of Modern Governments*, ed. A. Vidich and R.

Glassman (Beverly Hills, Calif.: Sage Publications 1979), pp. 17–48.

30. See A. Giddens's discussion of Habermas's work in *The Return of Grand Theory in the Human Sciences*, ed. Q. Skinner (Cambridge: Cambridge University Press, 1985), pp. 121–140; J. Habermas, *Legitimation Crisis*, trans. T. McCarthy (Boston: Beacon Press, 1975), pp. 26–27.

31. *Congressional Quarterly*, 10 December 1977, p. 2555.

32. House Subcommittee on Oversight and Investigations, *Hearings: Nuclear Waste Management and Disposal*, 95th Cong., 1st sess., 1977, 28, 101. In 1977, however, there was still the question of whether spent fuel would be considered waste. Heretofore it had been assumed that there would be spent fuel reprocessing to recover the uranium and plutonium after which the federal government would be responsible for disposing the high-level wastes from reprocessing. Nevertheless, ERDA and DOE went ahead with plans for a repository which could accommodate spent fuel, at one point proposing use of the Waste Isolation Pilot Plant in New Mexico for this purpose.

33. House Subcommittee on Oversight and Investigations, *Hearings*, 1977, p. 76. Please note that the term *spokesman* is being used accurately. In reviewing ten years of congressional hearings on nuclear waste policy, I did not encounter a spokeswoman for a nuclear industry.

34. Senate Subcommittee on Nuclear Regulation, *Hearings: Nuclear Waste Management*, 95th Cong., 2d sess., 1978, 122–27.

35. Senate Subcommittee on Energy, Nuclear Proliferation, and Federal Services, *Hearings: Report of the Interagency Review Group on Nuclear Waste Management*, 96th Cong., 1st sess., 1979, 1–3.

36. Senate Subcommittee on Nuclear Regulation, *Hearings*, 1978, p. 3.

37. Senate Subcommittee on Science, Technology, and Space, *Hearings: Nuclear Waste Disposal*, 95th Cong., 2d sess., 1978, 143.

38. See testimony by spokespersons from Nevada, Utah, Maine, Oregon, Massachusetts, National Governors Association, Sierra Club, Friends of the Earth, and the Office of Technology Assessment, in House Subcommittee on Energy Conservation and Power, *Hearings: Nuclear Waste Disposal Policy*, 97th Cong., 2d sess., 1982. DOE's plans for accelerating development were described in House Subcommittee on Energy and the Environment, *Hearings*, 1982, 62–74.

39. House Subcommittee on Energy Conservation and Power, *Hearings*, 1982, p. 356.

40. Ibid., 547.

41. Ibid., 550.

42. House Subcommittee on Energy and Power, *Hearings: Spent Fuel Storage and Disposal*, 96th Cong., 1st sess., 1979, 286–89.

43. See Cook, "Nuclear Follies"; S. Fenn, *America's Electric Utilities: Under Siege and in Transition* (New York: Praeger, 1984); Office of Technology Assessment, *Nuclear Power*; Flavin, *Nuclear Power*; Rudolph and Ridley, *Power Struggle*, pp. 144–79; J. Campbell, *Collapse of an Industry: Nuclear Power and the Contradictions of U.S. Policy* (Ithaca, N.Y.: Cornell University Press, 1988).

44. U.S. Office of Technology Assessment, *Nuclear Power*, p. 32.

45. J. Hewlett, *Investor Perceptions of Nuclear Power* (Washington, D.C.: Energy Information Administration, 1984).

46. Cook, "Nuclear Follies," pp. 88–90. Even in an environment of high interest rates, uncertain demand, and regulatory revisions, some utilities brought profitable nuclear power plants on line. Utilities with relatively little experience building nuclear power plants had the worst financial record.

47. House Subcommittee on Energy and Power, *Hearings*, 1979, 300.

48. Ibid., 152–58.

49. House Subcommittee on Energy and the Environment, *Hearings*, 1980, 9. In another attempt to use incentives, a group of United States' and Germany's nuclear industries offered Namibia eight hundred million dollars a year if it would host a radioactive waste disposal site. See *Nuclear Waste News*, 1 November 1984, p. 259.

50. A conclusion even reached by some who favor restoration of the nuclear option. See, for example, J. Morone and E. Woodhouse, *The Demise of Nuclear Energy? Lessons for Democratic Control of Technology* (New Haven: Yale University Press, 1988), chap. 4.

CHAPTER 4. *Reassertion or Reconstruction*

1. House Subcommittee on Energy Research and Production, *Hearings: Nuclear Waste Management Comprehensive Legis-*

lation—Bouquard/Lujan Proposal, 97th Cong., 1st sess., 1981, 93.

2. Department of Energy, *Backgrounder: Studies of Alternative Methods of Radioactive Waste Disposal* (Washington, D.C., 1986). The options of sub-seabed disposal and shipping waste to a Pacific island never completely disappeared as politicians from Nevada periodically mentioned them as alternatives to a repository at Yucca Mountain.

3. Senate Committee on Energy and Natural Resources and Committee on Environment and Public Works, *Joint Hearings: Nuclear Waste Disposal,* 97th Cong., 1st sess., 1981, 603. See also D. Metlay, "Radioactive Waste Management Policy-making," in Office of Technology Assessment, *Managing the Nation's Commercial High-level Radioactive Waste,* (Washington, D.C., 1985), pp. 203–04.

4. House Subcommittee on Energy Conservation and Power, *Hearings: Nuclear Waste Disposal Policy,* 97th Cong., 2d sess., 1982, 260–89. In response to the subcommittee's questions, DOE's written response included a year-by-year breakdown of siting activities.

5. For listings of data bases and bibliographic sources, see Department of Energy, *Information Services Directory* (Washington, D.C., 1986).

6. G. Clarfield and W. Wiecek, *Nuclear America: Military and Civilian Nuclear Power in the U.S., 1940–1980* (New York: Harper & Row, 1984), pp. 369, 381.

7. S. Hilgartner, R. Bell and R. O'Connor, *Nukespeak: The Selling of Nuclear Technology in America* (New York: Penguin Books, 1983).

8. Senate Subcommittee on Nuclear Regulation, *Hearings: Nuclear Waste Disposal,* 96th Cong., 2d sess., 1980, 301–03.

9. House Subcommittee on Energy Conservation and Power, *Hearings,* 1982, 245.

10. Senate Committees on Energy and Natural Resources and the Environment and Public Works, *Joint Hearings,* 1981, 226–27, 229, 234, 388.

11. N. Moss, *The Politics of Uranium* (New York: Universe Books, 1982), pp. 114–19.

12. Numerous examples of criticisms by the Environmental Policy Institute and Natural Resources Defense Council of DOE site selection procedures between 1979 and 1981 may be found

in: Senate Committee on Governmental Affairs, *Hearings: Nuclear Waste Management Reorganization Act of 1979*, 96th Cong., 1st sess., 1980, 128–46; House Subcommittee on Energy Research and Production, *Hearings: H.R. 7418—Nuclear Waste Research Development and Demonstration Act of 1980*, 96th Cong., 2d sess., 1980, 75–79; House Subcommittee on Energy and Power, *Hearings: Nuclear Waste Disposal*, 96th Cong., 2d sess., 1980, 236–40.

13. Senate Committees on Energy and Natural Resources and the Environment and Public Works, *Joint Hearings*, 1981, 323.

14. House Subcommittee on Energy and the Environment, *Hearings: High-level Nuclear Waste Management*, 97th Cong., 2d sess., 1982, 38.

15. Senate Committees on Energy and Natural Resources and the Environment and Public Works, *Joint Hearings*, 1981, 572–73. This repeated fear of delay is interesting in that, years before, the AEC had voiced a similar concern when asked about the need to study a second back-up site in addition to the Lyons, Kansas, site. In any event, DOE would eventually have to delay the opening of a repository by five years or more due to problems with its quality assurance program and other aspects of the program.

16. House Subcommittee on Energy and the Environment, *Hearings: Radioactive Waste Legislation*, 97th Cong., 1st sess., 1981, 334–47.

17. A selection of proposals for alternative institutional designs may be found in the following materials: For discussion by Senator Charles Percy of Illinois and others on DOE's lack of credibility and the need for new institutional arrangements, see 28 July 1980, *Congressional Record* 96th Cong., 2d sess., 19982–984. For a statement by Morris Udall on the need for a nuclear safety board and state participation, see House Subcommittee on Energy and the Environment, *Hearings: To Amend the Atomic Energy Act of 1954*, 96th Cong., 2d sess., 1980, 850–52, 881–82. In the debate on HR 6390, which would amend the Atomic Energy Act of 1954, proposals were made for altering the nuclear establishment by creating an independent nuclear safety board, extending the power of the Environmental Protection Agency over DOE, revamping Nuclear Regulatory Commission regulations, and delaying the construction of additional nuclear reactors. Also see House Committee on Interior and Insular Affairs, *Report: Atomic Energy Act Amend-*

ments of 1980, 96 Cong. 2d sess., H. Rept. 96–1382, pt. 2. The Interagency Review Group proposed various reforms. See House Subcommittee on Energy and the Environment, *Hearings: Nuclear Waste Management,* 96th Cong., 1st sess., 1979. President Carter advocated a state planning council which would recommend procedural mechanisms for subsequent nuclear waste planning, review the development of comprehensive waste management plans, examine site selection criteria, and evaluate guidelines and procedures for site characterization and site selection. Eighteen members would comprise the council—fourteen would be elected officials from state, local, and tribal governments. See *Communication from the President: State Planning Council on Radioactive Waste Management Act,* 96th Cong. 2d. sess., 1980, H. Doc. 96-339. Proposals from the National Governors' Association are described in many of the congressional hearings held at this time. For example, see Senate Subcommittee on Nuclear Regulation, *Hearings: Nuclear Waste Disposal,* 96th Cong., 2d sess., 1980, 415–19. For discussion of a proposed approach to site selection which would include broad state and federal participation, see B. Solomon and D. Cameron, "Nuclear Waste Repository Siting: An Alternative Approach," *Energy Policy* (1985): 564–580.

18. House Subcommittee on Energy Research and Production, *Hearings,* 1981, 269–70. Also see OTA's testimony before the Senate Committees on Energy and Natural Resources and the Environment and Public Works, *Joint Hearings,* 1981, 259–98.

19. House Subcommittee on Energy Research and Production, *Hearings,* 1981, 275.

20. Ibid., 656–59.

21. House Subcommittee on Energy Conservation and Power, *Hearings,* 1982, 600.

22. House Subcommittee on Energy Conservation and Power, *Hearings: Nuclear Waste Disposal Policy,* 97th Cong., 2d sess., 1982, 249.

23. Ibid., 389

24. House Committee on Interior and Insular Affairs, *Report: Atomic Energy Act Amendments of 1980,* 96th Cong., 2d sess., 1980, H. Rept. 96-1382, pt. 2: 41–45.

25. House Subcommittee on Energy and the Environment, *Hearings,* 1981, 485–86.

26. J. Christofferson, *History: Federal Nuclear Waste Disposal Program in Paradox Basin, Utah,* (Salt Lake City, Utah: State of Utah High-Level Nuclear Waste Office, 1985), p. 2.

27. House Subcommittee on Energy Research and Production, *Hearings,* 1981, 214.

28. House Subcommittee on Energy Research and Production, *Hearings,* 1981, 331–32.

29. *Radioactive Waste Management: Message from the President Transmitting a Report on His Proposals for a Comprehensive Radioactive Waste Management Program,* 96th Cong., 2d sess., 1980, H. Doc. 96-266, 2.

30. Representatives of nuclear industries and utilities blamed institutional instability for lack of a waste disposal solution. See, for example, the testimony of Betram Wolfe, vice-president of General Electric and spokesman for the Atomic Industrial Forum, House Subcommittee on Energy and the Environment, *Hearings,* 1981, 570–90.

31. Senate Committees on Energy and Natural Resources and the Environment and Public Works, *Joint Hearings,* 1981, 566.

32. Many members of the utility industry concluded that nothing less than a fully operational repository was needed to answer California's ban on new nuclear power plants. See, for example, the testimony of J. Edward Howard, vice-president of Boston Edison and chairman of the steering committee of the Utility Nuclear Waste Management Group, in House Subcommittee on Energy Research and Production, *Hearings,* 1980, 98–110. In general, the Reagan administration DOE had been charged with investigating ways to promote the development of new nuclear power plants. See *Communication from the President of the United States: Nuclear Waste Disposal,* 97th Cong., 2d sess., 1982, H. Doc. 97-172. In 1984, DOE established a task force to investigate where DOE could intervene in the proceedings of state regulatory agencies to promote the completion of nuclear power plants. See *Inside Energy/ with Federal Lands,* 22 October 1984, pp. 1–2.

33. House Subcommittee on Energy Research and Production, *Hearings,* 1981, 1–2.

34. House Subcommittee on Energy Research and Production, *Hearings,* 1981, 93. See testimony on behalf of the Utility Waste Management Group, Edison Electric Institute, and the American Nuclear Energy Council.

35. Congressman Santini of Nevada argued in favor of an absolute state veto. As an alternative, he argued for requiring passage of a joint resolution of Congress to override a state veto which would put the burden of proof on DOE to defend its siting decision—and not on the states to demonstrate the inadequacy of that decision. See House Subcommittee on Energy Conservation and Power, *Hearings*, 1982, 222–25. The National Governors Association had been debating the issues of concurrence and the state veto since the 1970s. Speaking for NGA in testimony before Congress, Governor Evans of Idaho advocated state participation rather than federal preemption. See Senate Committee on Energy and Natural Resources, *Hearings: Nuclear Waste and Facility Siting Policy,* 96th Cong., 1st sess., pt. 1, 1979, 41–45. In 1982, its spokesman, Governor Robert List of Nevada, described NGA policy as favoring a state veto with the possibility of an override by both houses of Congress. See House Subcommittee on Energy Conservation and Power, *Hearings*, 1982, 395–404.

36. Senate Subcommittee on Nuclear Regulation, *Hearings: National Nuclear Waste Policy Act of 1981*, 97th Cong., 1st sess., 1981, 163.

37. Senate Committee on Governmental Affairs, *Hearings*, 1980, 203–06.

38. House Subcommittee on Energy and the Environment, *Hearings: To Amend the Atomic Energy Act of 1954*, 96th Cong., 2d sess., 1980, 621. Also see the complete testimony of the Edison Electric Institute at this hearing.

39. Senate Committees on Energy and Natural Resources and the Environment and Public Works, *Joint Hearings*, 1981, 221.

40. Senate Subcommittee on Nuclear Regulation, *Hearings*, 1981, 159.

41. House Subcommittee on Energy Research and Production, *Hearings*, 1981, 53–64. See the complete testimony of David Berick, Environmental Policy Center.

42. House Subcommittee on Energy and the Environment, *Hearings*, 97th Cong., 1st sess., 1981, 288.

43. Ibid., 557.

44. House Subcommittee on Energy Research and Production, *Hearings*, 1981, 337.

45. Senate Committees on Energy and Natural Resources and the Environment and Public Works, *Joint Hearings*, 1981, 338–41.

46. Senate Subcommittee on Nuclear Regulation, *Hearings*, 1981, 46–47.

47. House Subcommittee on Energy Research and Production, *Hearings*, 1981, 302.

CHAPTER 5. *The Failure of the 1982 Nuclear Waste Policy Act*

1. Summaries of the debate in the House of Representatives on nuclear waste legislation are found in the *Congressional Quarterly: Weekly Report* 40 (1982): 1061–63, 1466, 1548, 1903–04, 2366, 2422, 2951–52, 3103–06. The Senate debate may be similarly followed and documented.

2. *Nuclear Waste Policy Act of 1982*, Public Law 97–425, sec. 111(b)(1), hereafter cited as NWPA.

3. NWPA, sec. 111(a)(1–4).

4. NWPA, sec. 112(b)(2–3).

5. Over the years, DOE opposed application of civilian safety, environmental, and health standards to its weapons facilities and operations. For an example of how this opposition delayed the clean-up of nuclear waste sites at weapons facilities, see E. Marshall, "The Buried Cost of the Savannah River Plant," *Science* 233 (1986): 613–-15. It is important to recognize that DOE facilities are actually operated by private contractors. DOE intended to turn the repository program over to such a contractor (Bechtel) until the contract award was challenged in court. The adequacy of DOE's supervision of its contractors, and charges that in the past DOE ordered its contractors to ignore environmental laws, continue to be issues in the case of the Rocky Flats weapons plant near Denver.

6. House Subcommittee on Energy and the Environment, *Hearings: High-level Nuclear Waste Management*, 97th Cong., 2d sess., 1982, 2.

7. Ibid., 60.

8. Ibid., 61.

9. Senate Committee on Environment and Public Works, *Report: National Nuclear Waste Regulation and Control Act of 1980*, 96th Cong., 2d sess., 1980, S. Rept. 96-871, 8.

10. For sources and pre-NWPA reports on this issue, see C. Pollock, "The Closing Act: Decommissioning Nuclear Power Plants," *Environment* 28, no. 2 (1986): 10–15, 33–36.

11. NWPA, sec. 304.

12. NWPA, sec. 121(a).

13. General Accounting Office, *The Energy Department's Office of Environment Does Not Have a Large Role in Decision-Making,* EMD-80-50 (Washington, D.C., 1980).

14. *Congressional Quarterly,* 10 December 1977, p. 2555.

15. NWPA, sec. 117(b).

16. Ibid., sec. 117(c).

17. Ibid., sec. 117(c)(8).

18. Ibid., sec. 117(c)(11).

19. *Nuclear News,* May 1986, pp. 115–16.

20. NWPA, sec. 8(a).

21. Ibid., sec. 9.

22. Ibid., sec. 303.

23. Department of Energy, *Draft Report to the Secretary of Energy on the Conclusions and Recommendations of the Advisory Panel on Alternative Means of Financing and Managing (AMFM) Radioactive Waste Management Facilities* (Washington, D.C., 1984), ch. 12, pp. 2–3.

24. *Nuclear News* 4, no. 37 (1984): 229.

25. Attempts to limit state participation in repository siting can be found throughout the debate on various nuclear waste policy acts. Criticisms of congressional attempts to limit state participation and override state regulations can be found in: House Subcommittee on Energy Conservation and Power, *Hearings,* 1982, 355–57, 395–404, 541–56.

26. NWPA, sec. 222.

27. House Subcommittee on Energy and the Environment, *Hearings: Radioactive Waste Legislation,* 97th Cong., 1st sess., 1981, 207, 241–42.

28. General Accounting Office to Rep. Richard Ottinger, Chairman, House Subcommittee on Energy Conservation and Power, 27 January 1984, "Re: DOE Needs to Evaluate Fully the Waste Management Effects of Extending the Useful Life of Nuclear Fuel." Letter.

29. Ibid.

30. Ibid.

31. Department of Energy, *Draft Mission Plan Amendment* (Washington, D.C., 1987), pp. 52–53, and *Mission Plan for the Civilian Radioactive Waste Management Program* (Washington, D.C., 1985), vol. 1, pp. 377–83.

32. S. Meyers, "Institutional Aspects of Radioactive Waste Management," presented to the Radioactive Waste Conference, University of Wisconsin-Madison, 24 April 1981. The concept of a test and evaluation facility separate from the repository was not discussed in DOE's Mission Plan. Nevertheless, it was an option that could be held in reserve.

33. Department of Energy, *Mission Plan*, vol. 1, pp. 47–48.

34. NWPA, sec. 113(b–c).

35. NWPA, sec. 112(a).

36. House Subcommittee on Energy Conservation and Power, *Hearings*, 1982, 1–3. A federal interim storage facility (as defined by subtitle IB of the NWPA) became an unrealized option as DOE turned its attention to assisting utilities in the development and licensing of "dry cask storage" for spent fuel at individual reactor sites. In part, dry cask storage was desirable because of a shortage of reliable casks for transporting spent fuel to a central, interim storage facility. Shipments in old casks of questionable reliability would have subjected the program to even more intense public and congressional scrutiny than was focused on shipments of damaged fuel rods from the Three Mile Island reactor. Eventually, however, DOE proposed a monitored retrievable storage (MRS) facility that, pending completion of a repository, would serve many of the same functions as an interim storage facility (NWPA, subtitle IC).

37. House Subcommittee on Energy and the Environment, *Hearings*, 1981, 531–32; House Subcommittee on Energy Conservation and Power, *Hearings*, 1982, 244.

38. NWPA, sec. 131(a)(1–3).

39. Ibid., sec. 131(a)(1–3).

40. Department of Energy, *Bulletin*, September 1986, pp. 2, 4.

41. NWPA, sec. 302(a)(5)(A–B).

42. *The Radioactive Exchange* 3, no. 14 (18 September 1984): 3.

43. *Inside Energy/With Federal Lands*, 22 October 1984, pp. 1–2.

44. NWPA, sec. 137(a)(2).

45. House Subcommittee on Energy and the Environment, *Hearings*, 1981, 540.

46. *Nuclear Waste News* 4, no. 44 (15 November 1984): 272.

47. M. Resnikoff, *The Next Nuclear Gamble: Transportation and Storage of Nuclear Waste* (New York: Council on Economic Priorities, 1983). This book references numerous government and

other documents that detail safety studies and accident statistics, descriptions of highway safety problems, and the inadequacy of federal inspection programs.

48. NWPA, sec. 9.

49. Ibid., sec. 117(c)(7).

50. House Subcommittee on Energy Conservation and Power, *Hearings*, 1982, 234–35. This theme is developed in chapter 7.

51. *Congressional Quarterly: Weekly Report* 40 (1982): 262.

52. This position was outlined by DOE's Energy Research Advisory Board. *Inside Energy/With Federal Lands*, 5 November 1984, pp. 1–3.

53. House Subcommittee on Energy and the Environment, *Hearings*, 1981, 557.

CHAPTER 6. *Technological Solutions*

1. A. Weinberg, "Social Institutions and Nuclear Energy," *Science* 177, no. 4043 (1972): 34.

2. Human Interference Task Force, *Communication Measures to Bridge 10 Millennia* (Columbus, Ohio: Battelle Memorial Institute Office of Nuclear Waste Isolation, 1984). See also Human Interference Task Force, *Reducing the Likelihood of Future Human Activities that Could Affect Geologic High-Level Waste Repositories*, BMI/ONWI-537 (Columbus, Ohio: Battelle Memorial Institute Office of Nuclear Waste Isolation, 1984).

3. *Salt Lake Tribune*, 12 October 1984, p. B-1.

4. A. Dravo, "The Politics of Federal-State Relations," presented at the Waste Management '86 Conference, Tucson, Arizona, 1986.

5. Ibid.

6. Plans to explore salt sites in Ohio were dropped, according to critics, because of political pressure. See House Subcommittee on Energy Conservation and Power, *Hearings: Nuclear Waste Disposal Policy,*, 97th Cong., 2d sess., 1982, 545.

7. J. Christofferson, *History: Federal Nuclear Waste Disposal Program in Paradox Basin, Utah* (Salt Lake City, Utah: Utah High-Level Nuclear Waste Office Contract Report #85-0111, 1985). This report, written by a former assistant to Utah governor Scott Matheson, documents conflicts between the state and DOE over the repository siting program between 1976 and 1984.

8. Ibid., pp. 40–41.

9. S. Jasanoff and D. Nelkin, "Science, Technology and the Limits of Judicial Competence," in *Resolving Locational Conflict*, ed. R. Lake (New Brunswick, N.J.: Rutgers University Center for Urban Policy Research, 1987), pp. 60–71.

10. For descriptions of these, and numerous other, attempts to manipulate public access to information critical of nuclear technologies, see S. Hilgartner, R. Bell, and R. O'Connor, *Nukespeak: The Selling of Nuclear Technology in America* (New York: Penguin Books, 1983).

11. In 1987, two congressmen, Ron Wyden of Oregon and Al Swift of Washington, asked the General Accounting Office to investigate charges that DOE's chief contractor at Hanford concealed information on groundwater contamination which could eliminate the site from consideration as a nuclear waste repository. A previous investigation by DOE's inspector general had determined there was insufficient evidence to substantiate the charges. However, the congressmen, pointing to internal memos overlooked in that investigation, requested a GAO investigation. DOE and its contractors allegedly had threatened to fire employees who communicated with the Nuclear Regulatory Commission without management approval, and who otherwise publicly discussed conditions at Hanford which could make it unsuitable for a repository. A Rockwell employee (Rockwell was DOE's chief contractor at Hanford at the time) was allegedly dismissed for talking with NRC and discussing the results of a 1975 study that found radioactive contamination—specifically Iodine-129—traveled much faster in Hanford's aquifers than previously estimated. Accelerated groundwater travel times increased the probability of a repository at Hanford polluting the Columbia River. See: *Inside Energy/With Federal Lands*, 24 August 1987, p. 3; General Accounting Office, *Nuclear Waste: DOE's Handling of Hanford Reservation Iodine Information*, GAO/RCED-88-158 (Washington, D.C., 1988).

12. G. Jacob, "State Technical Review of the High-Level Waste Program," *Proceedings: Waste Management '86* (Tucson, Ariz.: University of Arizona, 1986), pp. 69–71.

13. G. Benveniste, *The Politics of Expertise* (Berkeley: Glendessary Press, 1972), p. 202.

14. M. Edelman, *The Symbolic Uses of Politics*, 2d ed. (Urbana, Ill.: University of Illinois Press, 1985), p. 3.

15. *Potential Impacts of Characterization, Siting, and Construction of a High Level Nuclear Waste Repository in Southern Utah on National Parks: Summary of a Conference,* ed. R. Schreyer et al. (Logan, Utah: Utah State University Department of Forest Resources, 1984).

16. D. Torgerson, "Contextual Orientation in Policy Analysis: The Contribution of Harold D. Lasswell," *Policy Sciences* 18 (1985): 241–261.

17. Benveniste, *Politics of Expertise,* p. 193.

18. S. Del Sesto, *Science, Politics, and Controversy: Civilian Nuclear Power in the U.S., 1946–1974* (Boulder, Colo.: Westview Press, 1979), pp. 181–208.

19. S. Hadden, J. Chiles, P. Anaejionu, and K. Cerny, *High Level Nuclear Waste Disposal: Information Exchange and Conflict Resolution* (Austin, Tex.: Texas Energy and Natural Resources Advisory Council, 1981).

20. Department of Energy, *Mission Plan for the Civilian Radioactive Waste Management Program* (Washington, D.C., 1985), vol. 1, pp. 129–37; Department of Energy, *Mission Plan Amendment* (Washington, D.C., 1987), pp. 27–32.

21. Research on public attitudes toward nuclear power has shown that technical information may have little to do with shifting support for nuclear power. See J. Kuklinski, D. Metlay, and W. Kay, "Citizen Knowledge and Choices on the Complex Issue of Nuclear Energy," *American Journal of Political Science* 26, no. 4 (1982): 615–42; S. Rothman and S. Lichter, "Elites in Conflict: Nuclear Energy, Ideology, and the Perception of Risk," *Journal of Contemporary Studies* 8, no. 3 (1985): 23–44; J. van der Pligt, J. Eiser, and R. Spears, "Public Attitudes to Nuclear Energy," *Energy Policy* 12, no. 3 (1984): 302–05.

22. D. Nelkin, "Science, Technology, and Political Conflict: Analyzing the Issues," in *Controversy: The Politics of Technical Decisions,* 2d ed., ed. D. Nelkin (Beverly Hills, Calif.: Sage Publications, 1984), pp. 9–24.

CHAPTER 7. *The Political Uses of Location*

1. Department of Energy, Press Release R-86-065, 15 April 1986, Office of the Press Secretary, Washington, D.C.

2. J. Wolpert, "Departures From the Usual Environment in Locational Analysis," *Annals of the Association of American Geographers* 60, no. 2 (1970): 222.

3. Ibid., 224.

4. Historical background can be found in virtually any book or collection of articles cited here. From a political-geographic point of view, one of the more interesting accounts can be found in J. Seley, *The Politics of Public Facility Planning* (Lexington, Mass.: D. C. Heath & Co., 1983).

5. G. Rochlin, *Plutonium, Power and Politics; International Arrangements For the Disposition of Spent Nuclear Fuel* (Berkeley and Los Angeles: University of California Press, 1979).

6. In 1980 John Deutch, DOE under-secretary, said that DOE was already characterizing three sites in detail. See House Subcommittee on Energy and the Environment, *Hearings: To Amend the Atomic Energy Act of 1954*, 96th Cong., 2d sess., 1980, 116–17. For additional information on DOE's investment in potential repository sites previous to the 1982 NWPA, see also House Subcommittee on Energy and Power, *Hearings: Nuclear Waste Disposal*, 96th Cong., 2d sess., 1980, 209–18. For a complete list of DOE repository site- related work completed before the 1982 NWPA, see House Subcommittee on Energy Conservation and Power, *Hearings: Nuclear Waste Disposal Policy*, 97th Cong., 2d sess., 1982, 260–91.

7. The U.S. Air Force learned this lesson in Nevada and Utah. Usually very accommodating toward military programs, these states opposed the original MX missile "shell-game" proposal, pointing to the air force's arrogant insensitivity to local concerns and threats to the stability of the dominant local culture.

8. A. Kirby, *The Politics of Location* (London: Methuen, 1982), p. 5.

9. A.G. Mojitabai, *Blessed Assurance: At Home with the Bomb in Amarillo, Texas* (Boston: Houghton Mifflin Co., 1986).

10. *San Juan Record*, 11 July 1984.

11. *San Juan Record*, 27 June 1984.

12. DOE paid more attention to two sites in Mississippi. In 1978, Governor Edwin Edwards of Louisiana and John O'Leary, deputy secretary of energy, signed an agreement that offered Louisiana, in return for accepting the Strategic Petroleum Reserve, assurances that DOE would not construct a high-level waste repository if the state objected to such a facility. Donald Barlett and James

Steele, reporters for the *Philadelphia Inquirer*, quoted Ronald Reagan during the 1980 presidential election as saying that under his administration the federal government would live up to its agreement not to construct a nuclear waste repository in Louisiana. See D. Barlett and J. Steele, *Forevermore: Nuclear Waste in America* (New York: Norton, 1984), p. 159. The legal status of the agreement was doubtful, especially following passage of the NWPA, but went unchallenged. In 1987, Senator J. Bennett Johnston of Louisiana crafted revisions to the site selection process which effectively eliminated all sites except for Yucca Mountain, Nevada, from consideration. Characterization of sites in Washington and Texas, and the second round repository program, were terminated.

13. Department of Energy, *Environmental Assessment Overview: Richton Dome Site Mississippi* (Washington, D.C., 1986), pp. 14–15.

14. S. Hansen, "State Innovation and Nuclear Waste Disposal: Consultation and Concurrence in Michigan and Mississippi," presented at Waste Management '81, Tucson, Arizona, 23–26 February 1981.

15. *Dallas Morning News*, 17 October 1984.

16. See, for example, R. Johnston, *Geography and the State: An Essay in Political Geography* (New York: St. Martin's Press, 1982), pp. 1–10.

17. *Nuclear Waste Policy Act of 1982*, Public Law 97-425, sec. 115–17.

18. General Accounting Office, *Status of the DOE's Implementation of the NWPA of 1982*, GAO/RCED-85-116 (Washington D.C., 1985), p. 21.

19. These observations are based on the author's experience while working in the state of Utah's High-Level Nuclear Waste Office. Both states were active on the Western Interstate Energy Board's High-Level Waste Committee which considered regional transportation issues.

20. Department of Energy, *Transportation Institutional Plan* (Washington, D.C., 1986), p. A-96.

21. Nuclear Regulatory Commission, *Public Information Circular for Shipments of Irradiated Reactor Fuel*, NUREG-0725 (Washington, D.C., 1984). See also Department of Energy, *Transportation Institutional Plan* and numerous reports of these agencies cited in M. Resnikoff, *The Next Nuclear Gamble: Transportation*

and Storage of Nuclear Waste (New York: Council on Economic Priorities, 1983).

22. Department of Energy, *Environmental Assessment: Hanford Washington* (Washington, D.C., 1986), and *Environmental Assessment: Deaf Smith Texas* (Washington, D.C., 1986).

23. The Utah Geologic and Mineral Survey publicized its own assessment of repository sites in which they pointed to Nevada as being a better site than Utah. "Area in Southern Nevada best for nuclear waste." *Salt Lake Tribune,* 3 October 1984, sec. B, p. 2.

24. A point made by E. Bardach, *The Implementation Game: What Happens After a Bill Becomes a Law* (Cambridge, Mass.: MIT Press, 1977).

25. A form of power discussed in S. Lukes, *Power: A Radical View* (London: Macmillan Press Ltd., 1974); G. Benveniste, *The Politics of Expertise* (Berkeley: Glendessary Press, 1972); J. Gaventa, *Power and Powerlessness* (Urbana, Ill.: University of Illinois Press, 1980).

26. Gaventa, *Power and Powerlessness,* p. 42.

27. For a discussion of the geographic expression of externalities and risk, see D. Ziegler, J. Johnson, Jr., S. Brunn, *Technological Hazards* (Washington, D.C.: Association of American Geographers, 1983), pp. 35–38.

28. As another example, see M. Crenson, *The Un-Politics of Air Pollution: A Study of Non-Decisionmaking in the Cities* (Baltimore: Johns Hopkins University Press, 1971).

29. For example, see how transportation impacts are narrowly defined in the environmental assessment for one of the Utah sites. Department of Energy, *Environmental Assessment: Davis Canyon, Utah* (Washington, D.C., 1986).

30. Department of Energy, *Environmental and Socioeconomic Considerations of Locating a Nuclear Waste Repository near Canyonlands National Park, Utah* (Washington, D.C., 1984).

31. Ibid., pp. 4, 22, 246.

32. Colorado Governor Richard Lamm and State Senator Tom Glass, to Nunzio Palladino, chairman, Nuclear Regulatory Commission, 15 August 1984. Photocopy.

33. DOE Office of Civilian Radioactive Waste Management, *Bulletin* (December 1986): 5.

34. General Accounting Office, *The Nation's Nuclear Waste:*

Proposals for Organization and Siting, EMD-79-77 (Washington, D.C., 1979), pp. iv–v.
    35. Ibid, p. 18.

CHAPTER 8. *Conclusion*

1. S. Barrett and C. Fudge, "Examining the Policy-Action Relationship," in *Policy and Action: Essays on the Implementation of Public Policy,* ed. S. Barrett and C. Fudge (London: Methuen, 1981), pp. 3–32.
    2. *Inside Energy/With Federal Lands,* 20 February 1989, pp. 6–7.
    3. L. Carter, "Siting the Nuclear Waste Repository: Last Stand at Yucca Mountain," *Environment* 29, no. 8 (1987): 8–13, 26–32.
    4. Interview with Harry Swainston, deputy attorney general, State of Nevada, 1 August 1989. Also see: *Inside Energy/With Federal Lands,* 10 July 1989, p. 2.
    5. In addition to terminating work on other geologic formations in the United States, DOE terminated cooperative efforts on other waste disposal options. Work with the Swedish government on the long-term stability of repositories in crystalline rock was discontinued. At a cost of $3.2 million, DOE terminated work with Atomic Energy of Canada Ltd. at the underground research laboratory. See General Accounting Office, *Nuclear Waste: DOE Has Terminated Research Evaluating Crystalline Rock for A Repository,* GAO/RCED-89-148 (Washington, D.C., 1989).
    6. General Accounting Office, *Nuclear Waste: DOE's Handling of Hanford Iodine Information,* GAO/RCED-88-158 (Washington, D.C., 1988).
    7. General Accounting Office, *Nuclear Health and Safety: Stronger Oversight of Asbestos Control Needed at Hanford Tank Farms,* GAO/RCED-88-150 (Washington, D.C., 1988).
    8. *Radioactive Exchange,* 22 January 1988, pp. 1–4; 15 May 1988, p. 12; *Edison Electric Institute Washington Letter,* 22 January 1988, pp. 4–5; *Inside Energy/With Federal Lands,* 25 January 1988.
    9. *Nuclear Waste News,* 28 April 1988, pp. 135–36. See also: 29 September 1988, p. 308; 17 November 1988, p. 368; 13 July 1989, pp. 249–50; *Radioactive Exchange,* 15 May 1988, pp. 1, 7–8;

and Nuclear Regulatory Commission, *News Releases,* 11 July 1989 (NUREG/BR-0032, vol. 9, no. 28).

10. Employees of DOE's Office of Civilian Radioactive Waste Management rightly point out that the weapons and repository programs are separate. At the organizational level this is the case; but, many employees move freely between the civilian and weapons programs, taking temporary assignments with each as required. In any event, the public is unlikely to make such fine distinctions among programs when assessing the credibility, legitimacy, and integrity of the agency itself.

11. Remarks by James D. Watkins, Secretary of Energy, 27 June 1989, pp. 2–3. Accompanied by DOE Press Release R-89-068, "Watkins Announces Ten-Point Plan for Environmental Protection, Waste Management."

12. Ibid., pp. 1–2.

13. Ibid., pp. 5–11.

14. Ibid., p. 13.

15. See various reports in *NUEXCO* (a trade publication of the uranium industry) and *Inside Energy/With Federal Lands,* 6 March 1989.

16. Steven Kraft, director, Utility Nuclear Waste Management Group, quoted in an interview with *Inside Energy/With Federal Lands,* 14 November 1988, p. 3.

17. General Accounting Office, *Nuclear Health and Safety: Oversight at DOE's Nuclear Facilities Can Be Strengthened,* GAO/RCED-88-137 (Washington, D.C., 1988).

18. Summary in *Nuclear Waste News,* 21 July 1988, p. 233 of *The Nuclear Congress: How Congress Voted on Nuclear Power Issues, 1986–1988* (Washington, D.C.: Public Citizen, 1988).

BIBLIOGRAPHY

▼

Ackerman, B., and W. Hassler. *Clean Coal/Dirty Air*. New Haven: Yale University Press, 1982.

Alford, R., and R. Friedland. *Powers of Theory: Capitalism, the State and Democracy*. Cambridge: Cambridge University Press, 1985.

Arnold, R. *Congress and the Bureaucracy: A Theory of Influence*. New Haven: Yale University Press, 1979.

Bachrach, P., and M. Baratz. "Two Faces of Power." *American Political Science Review* 56 (1962): 947–52.

———. "Decisions and Nondecisions: An Analytic Framework." *American Political Science Review* 57 (1963): 632–42.

Bardach, E. *The Implementation Game: What Happens After a Bill Becomes a Law*. Cambridge, Mass.: MIT Press, 1977.

Barlett, D., and J. Steele. *Forevermore: Nuclear Waste in America*. New York: W. W. Norton, 1985.

Barrett, S., and C. Fudge. "Examining the Policy-Action Relationship." In *Policy and Action: Essays on the Implementation of Public Policy*, ed. S. Barrett and C. Fudge, 3-32. London: Methuen, 1981.

Bensman, J. "Max Weber's Concept of Legitimacy." In *Conflict and Control: Challenge to Legitimacy of Modern Governments*, ed. A. Vidich and R. Glassman, 17–48. Beverly Hills, Calif.: Sage Publication, 1979.

Benveniste, G. *The Politics of Expertise*. Berkeley: Glendessary Press, 1972.

Bishop, W., I. Hoos, N. Hilberry, D. Metlay, and R. Watson, eds. *Essays on Issues Relevant to the Regulation of Radioactive Waste Management*. Washington, D.C.: U.S. Nuclear Regulatory Commission, 1978.

Boddy, M. "Central-Local Government Relations: Theory and Practice." *Political Geography Quarterly* 2, no. 2 (1983): 119–38.

Burnett, A., and P. Taylor. *Political Studies from Spatial Perspectives.* New York: John Wiley & Sons, 1981.

Campbell, J. *Collapse of an Industry: Nuclear Power and the Contradictions of U.S. Policy.* Ithaca, N.Y.: Cornell University Press, 1988.

Carnes, S. "Confronting Complexity and Uncertainty: Implementation of Hazardous Waste Management Policy." In *Environmental Policy Implementation,* ed. D. Mann, 35–50. Lexington, Mass.: Lexington Books, 1982.

Carter, L. *Nuclear Imperatives and Public Trust: Dealing with Radioactive Waste.* Washington, D.C.: Resources for the Future, 1987.

———. "Siting the Nuclear Waste Repository: Last Stand at Yucca Mountain." *Environment* 29, no. 8 (1987): 8–13, 26–32.

Christofferson, J. *History: Federal Nuclear Waste Disposal Program in Paradox Basin, Utah.* Contract Report #85-0111. Salt Lake City, Utah: Utah High-Level Nuclear Waste Office, 1985.

Clarfield, G., and W. Wiecek. *Nuclear America: Military and Civilian Nuclear Power in the United States, 1940–1980.* New York: Harper & Row, 1984.

Colglazier, E., ed. *The Politics of Nuclear Waste.* New York: Pergamon Press, 1982.

Congressional Quarterly, 10 December 1977.

Congressional Quarterly, 18 March 1978.

Congressional Quarterly: Weekly Report 40 (1982): 261–62, 1061–63, 1466, 1548, 1903–04, 2366, 2422, 2951–52, 3103–06.

Cook, J. "Nuclear Follies." *Forbes,* 11 February 1985, pp 82–100.

Covello, V. "The Perception of Technological Risks: A Literature Review." *Technological Forecasting and Social Change* 23 (1983): 285-97.

Cox, K. "Residential Mobility, Neighborhood Activism and Neighborhood Problems." *Political Geography Quarterly* 2, no. 2 (1983): 99-118.

Crenson, M. *The Un-Politics of Air Pollution: A Study of Non-Decisionmaking in the Cities.* Baltimore: Johns Hopkins University Press, 1971.

Dallas Morning News, 17 October 1984.

Dear, M. "A Theory of the Local State." In *Political Studies from Spatial Perspectives,* ed. A. Burnett and P. Taylor, 183-200. New York: John Wiley & Sons, 1981.

Dear, M., and G. Clark. "The State and Geographic Process: A Critical Review." *Environment and Planning—A* 10 (1978): 173–83.

Del Sesto, S. *Science, Politics, and Controversy: Civilian Nuclear Power in the United States, 1946–1974.* Boulder, Colo.: Westview Press, 1979.

Downey, G. "Politics and Technology in Repository Siting: Military Versus Commercial Nuclear Wastes at WIPP, 1972–1985." *Technology in Society* 7 (1985): 47–75.

Dravo, A. "The Politics of Federal-State Relations." Presented at the Waste Management '86 Conference, Tucson Ariz., 1986.

Duncan, J., and D. Ley. "Structural Marxism and Human Geography: A Critical Assessment." *Annals Association of American Geographers* 72 (1982): 30–59.

Edelman, M. *The Symbolic Uses of Politics.* 2d ed. Urbana: University of Illinois Press, 1985.

Edison Electric Institute Washington Letter, 22 January 1988.

Fenn, S. *America's Electric Utilities: Under Siege and in Transition.* New York: Praeger, 1984.

Flavin, C. *Nuclear Power: The Market Test.* Washington, D.C.: Worldwatch Paper #57, 1983.

Ford, D. *The Cult of the Atom: The Secret Papers of the Atomic Energy Commission.* New York: Simon & Schuster, 1982.

Fuller, J. *We Almost Lost Detroit.* New York: Reader's Digest Press, 1975.

Gandara, A. *Electric Utility Decision-making and the Nuclear Option.* Santa Monica, Calif.: Rand Corporation, 1977.

Gaventa, J. *Power and Powerlessness: Quiescence and Rebellion in an Appalachian Valley.* Urbana: University of Illinois Press, 1980.

Gladwin, T. "Patterns of Environmental Conflict Over Industrial Facilities in the United States—1970–1980." *Natural Resources Journal* 20 (1980): 243–74.

Goldschmidt, B. *The Atomic Complex.* La Grange, Ill.: American Nuclear Society, 1982.

Habermas, J. *Legitimation Crisis.* Trans. T. McCarthy. Boston: Beacon Press, 1975.

Hadden, S., J. Chiles, P. Anaejionu, and K. Cerny. *High Level Nuclear Waste Disposal: Information Exchange and Conflict Resolution.* Austin, Tex.: Texas Energy and Natural Resources Advisory Council, 1981.

Hansen, S. "State Innovation and Nuclear Waste Disposal: Consultation and Concurrence in Michigan and Mississippi." Presented at Waste Management '81, Tucson, Ariz., 23–26 February 1981.

Hewlett, J. *Investor Perceptions of Nuclear Power.* Washington, D.C.: Energy Information Administration, 1984.

Hilgartner, S., R. Bell, and R. O'Connor. *Nukespeak: The Selling of Nuclear Technology in America.* New York: Penguin Books, 1983.

Huber, P. "The Bhopalization of American Tort Law." In *Hazards: Technology and Fairness,* ed. National Academy of Engineering, 89-110. Washington D.C.: National Academy Press, 1986.

Human Interference Task Force. *Communication Measures to Bridge Ten Millennia.* Columbus, Ohio: Battelle Memorial Institute Office of Nuclear Waste Isolation, 1984.

Human Interference Task Force. *Reducing the Likelihood of Future Human Activities That Could Affect Geologic High-Level Waste Repositories,* BMI/ONWI-537. Columbus, Ohio: Battelle Memorial Institute Office of Nuclear Waste Isolation, 1984.

Inside Energy/With Federal Lands, 22 October 1984.

———. 5 November 1984.

———. 25 January 1988.

———. 14 November 1988.

———. 20 February 1989.

———. 6 March 1989.

———. 10 July 1989.

Jacob, G. "Peer Review and State Technical Review." In *Proceedings: Waste Management '86,* 69-71. Tucson, Ariz.: University of Arizona, 1986.

Jasanoff, S., and D. Nelkin. "Science, Technology and the Limits of Judicial Competence." In *Resolving Locational Conflict,* ed. R. Lake, 60-71. New Brunswick, N.J.: Rutgers University Center for Urban Policy Research, 1987.

Johnson, J., and D. Zeigler. "Evacuation Planning for Technological Hazards." *Cities* 3, no. 2 (1986): 148–56.

Johnston, R. *Geography and the State: An Essay in Political Geography.* New York: St. Martin's Press, 1982.

Kirby, A. *The Politics of Location.* London: Methuen, 1982.

———. "State, Local State, Context and Spatiality." Institute of Behavioral Science, Boulder, Colo., 1987. Photocopy.

Kolko, G. *The Triumph of Conservatism.* New York: The Free Press, 1963.

Kuklinski, J., D. Metlay, and W. Kay. "Citizen Knowledge and Choices on the Complex Issue of Nuclear Energy." *American Journal of Political Science* 26, no. 4 (1982): 615–42.

Lamm, Governor Richard and State Senator Tom Glass. Letter to Nunzio Palladino, chairman, Nuclear Regulatory Commission, 15 August 1984. Photocopy.

Lanouette, W. "Atomic Energy 1945–85." *Wilson Quarterly 9*, no. 5 (1985): 91–131.

Lewis, R. *The Nuclear Power Rebellion: Citizens Versus the Atomic-industrial Establishment.* New York: Viking, 1972.

Lipschutz, R. *Radioactive Waste: Politics, Technology, and Risk.* Cambridge, Mass.: Ballinger, 1980.

Lovins, A. "Energy Strategy: The Road Not Taken." *Foreign Affairs* 55, no. 1 (1976): 65–96.

Lukes, S. *Power: A Radical View.* London: Macmillan, 1974.

Macgill, S. "Exploring the Similarities of Different Risks." *Environment and Planning—B* 10 (1983): 303–29.

Mann, D., ed. *Environmental Policy Implementation: Planning and Management Options and Their Consequences.* Lexington, Mass.: Lexington Books, 1982.

Marshall, E. "The Buried Cost of the Savannah River Plant." *Science* 233 (1986): 613–15.

Metlay, D. "History and Interpretation of Radioactive Waste Management in the United States." In *Essays on Issues Relevant to the Regulation of Radioactive Waste Management,* ed. W. Bishop et al., 1-19. Washington, D.C.: Nuclear Regulatory Commission, 1978.

———. "Radioactive Waste Management Policy-making." In *Managing the Nation's Commercial High-level Radioactive Waste.* U.S. Office of Technology Assessment. Washington, D.C.: U.S. Congress, 1985.

Metzger, H. *The Atomic Establishment.* New York: Simon & Schuster, 1972.

Meyers, S. "Institutional Aspects of Radioactive Waste Management." Presented to the Radioactive Waste Conference, University of Wisconsin-Madison, 24 April 1981.

Mojitabai, A. *Blessed Assurance: At Home with the Bomb in Amarillo, Texas.* Boston: Houghton Mifflin Co., 1986.

Morone, J. and E. Woodhouse. *The Demise of Nuclear Energy? Les-*

sons for Democratic Control of Technology. New Haven: Yale University Press, 1989.

Moss, N. *The Politics of Uranium.* New York: Universe Books, 1982.

Murauskas, G. and F. Shelley. "Local Political Responses to Nuclear Waste Disposal." *Cities* 3, no. 2 (1986): 157–62.

Nealey, S., and J. Herbert. "Public Attitudes Toward Radioactive Wastes." In *Too Hot to Handle?: Social and Policy Issues in the Management of Radioactive Wastes,* ed. C. Walker, L. Gould, and E. Woodhouse, 94–111. New Haven: Yale University Press, 1983.

Nelkin, D. "Science, Technology, and Political Conflict: Analyzing the Issues." In *Controversy: Politics of Technical Decisions,* 2d ed., ed. D. Nelkin, 9-24. Beverly Hills, Calif.: Sage Publications, 1984.

Nevada Nuclear Waste Newsletter (State of Nevada Nuclear Waste Project Office), April 1986.

Nuclear News 4, no. 37 (1984): 229.

Nuclear Waste News. 1 November 1984.

———. 15 November 1984.

———. 28 April 1988.

———. 21 July 1988.

———. 29 September 1988.

———. 17 November 1988.

———. 13 July 1989.

Nuclear Waste Policy Act of 1982, Public Law 97-425.

Overcast, T., and B. Schuknecht. *Federal Preemption of State and Local Nuclear Transportation Regulations.* Albuquerque: Sandia National Laboratories Transportation Technology Center, 1985.

Pollock, C. "The Closing Act: Decommissioning Nuclear Power Plants." *Environment* 28, no. 2 (1986): 10–15, 33–36.

Pressman, J., and A. Wildavsky. *Implementation.* 3rd ed. Berkeley: University of California Press, 1984.

Radioactive Exchange. 18 September 1984.

———. 22 January 1988.

———. 15 May 1988.

Resnikoff, M. *The Next Nuclear Gamble: Transportation and Storage of Nuclear Waste.* New York: Council on Economic Priorities, 1983.

Rochlin, G. *Plutonium, Power and Politics; International Ar-*

rangements for the Disposition of Spent Nuclear Fuel. Berkeley and Los Angeles: University of California Press, 1979.

Rothman, S., and S. Lichter. "Elites in Conflict: Nuclear Energy, Ideology, and the Perception of Risk." *Journal of Contemporary Studies* 8, no. 3 (1985): 23–44.

Rourke, F. *Bureaucracy, Politics, and Public Policy.* 3rd ed. Boston: Little, Brown & Co., 1984.

Rudolph, R., and S. Ridley. *Power Struggle: The Hundred-Year War Over Electricity.* New York: Harper & Row, 1986.

Salt Lake Tribune. 3 October 1984, sec. B, p. 2.

———. 12 October 1984, sec. B, p. 1.

Sandbach, F. *Environment, Ideology and Policy.* Oxford: Basil Blackwell, 1980.

San Juan Record. 27 June 1984.

———. 11 July 1984.

Schreyer, R., D. Williams, J. Kennedy, P. Mohai, and R. Nichols, eds. *Potential Impacts of Characterization, Siting, and Construction of a High Level Nuclear Waste Repository in Southern Utah on National Parks: Summary of a Conference.* Logan, Utah: Utah State University, 1984.

Seley, J. *The Politics of Public Facility Planning.* Lexington, Mass.: D.C. Heath & Co., 1983.

Shelley, F., and G. Murauskas. "Local Conflict, Local Autonomy, and Regional Concern: Siting Nuclear Waste in South Dakota." Delivered at the Eighty-first Annual Meeting of the Association of American Geographers, Detroit, Mich., April 1985.

Skinner, Q., ed. *The Return of Grand Theory in the Human Sciences.* Cambridge: Cambridge University Press, 1985.

Skowronek, S. *Building a New American State: The Expansion of National Administrative Capacities, 1877–1920.* Cambridge: Cambridge University Press, 1982.

Slovic, P., and B. Fischhoff. "How Safe Is Safe Enough?" In *Too Hot to Handle?: Social and Policy Issues in the Management of Radioactive Waste,* ed. C. Walker, L. Gould, and E. Woodhouse, 112–50. New Haven, Conn. Yale University Press, 1983.

Solomon, B., and D. Cameron. "Nuclear Waste Repository Siting: An Alternative Approach." *Energy Policy* (1985): 564–80.

Temples, J. "The Politics of Nuclear Power: A Subgovernment in Transition." *Political Science Quarterly* 95 (1980): 239–60.

Tierney, S. "The Nuclear Waste Disposal Controversy." In *Controversy: Politics of Technical Decisions*, 2d ed., ed. D. Nelkin, 91-110. Beverly Hills, Calif.: Sage Publications, 1984.

Torgerson, D. "Contextual Orientation in Policy Analysis: The Contribution of Harold D. Lasswell." *Policy Sciences* 18 (1985): 241–61.

U.S. Congress. House. *Radioactive Waste Management: Message from the President Transmitting a Report on His Proposals for a Comprehensive Radioactive Waste Management Program.* 96th Cong., 2d sess., 1980, H. Doc. 96-266.

———. *Communication from the President: State Planning Council on Radioactive Waste Management Act.* 96th Cong., 2d sess., 1980, H. Doc. 96-339.

———. *Communication from the President of the United States: Nuclear Waste Disposal—Initiatives on the Safe and Efficient Disposal of Nuclear Waste.* 97th Cong., 2d sess., 1982, H. Doc. 97-172.

———. Committee on Interior and Insular Affairs. *Report: Atomic Energy Act Amendments of 1980.* 96th Cong., 2d sess., 1980, H. Rept. 96-1382, pt. 2.

———. Committee on Science and Technology. *Nuclear Waste Research, Development, and Demonstration Act of 1980.* 96th Congress, 2d sess., 1980. H. Rept. 96-1156, pt.1.

———. Subcommittee on Energy Conservation and Power. *Hearings: Nuclear Waste Disposal Policy.* 97th Cong., 2d sess., 1982.

———. Subcommittee on Energy and the Environment. *Hearings: Nuclear Waste Disposal in Michigan.* 94th Cong., 2d sess., 1976.

———. Subcommittee on Energy and the Environment. *Hearings: Nuclear Waste Management.* 96th Cong., 1st sess., 1979.

———. Subcommittee on Energy and the Environment. *Hearings: To Amend the Atomic Energy Act of 1954.* 96th Cong., 2d sess., 1980.

———. Subcommittee on Energy and the Environment. *Hearings: Radioactive Waste Legislation.* 97th Cong., 1st sess., 1981.

———. Subcommittee on Energy and the Environment. *Hearings: High-level Nuclear Waste Management.* 97th Cong., 2d sess., 1982.

———. Subcommittee on Energy and Power. *Hearings: Spent Fuel Storage and Disposal.* 96th Cong., 1st sess., 1979.

————. Subcommittee on Energy and Power. *Hearings: Nuclear Waste Disposal.* 96th Cong., 2d sess., 1980.

————. Subcommittee on Energy Research and Production. *Hearings: H.R. 7418—Nuclear Waste Research Development and Demonstration Act of 1980.* 96th Cong., 2d sess., 1980.

————. Subcommittee on Energy Research and Production. *Hearings: Nuclear Waste Management Comprehensive Legislation—Bouquard/Lujan Proposal.* 97th Cong., 1st sess., 1981.

————. Subcommittee on Oversight and Investigations. *Hearings: Nuclear Waste Management and Disposal.* 95th Cong., 1st sess., 1977.

U.S. Congress. Senate. Committee on Energy and Natural Resources. *Hearings: Nuclear Waste and Facility Siting Policy.* 96th Cong., 1st sess., pt. 1, 1979.

————. Committee on Energy and Natural Resources and the Committee on Environment and Public Works. *Joint Hearings: Nuclear Waste Disposal.* 97th Cong., 1st sess., 1981.

————. Committee on Environment and Public Works. *Report: National Nuclear Waste Regulation and Control Act of 1980.* 96th Cong., 2d sess., 1980, S. Rept. 96-871.

————. Committee on Governmental Affairs. *Hearings: Nuclear Waste Management Reorganization Act of 1979.* 96th Cong., 1st sess., 1980.

————. Debate on Nuclear Waste Policy Act, S. 2189. *Congressional Record.* 96th Cong., 2d sess., 28 July 1980, pp. 19982–984.

————. Subcommittee on Energy, Nuclear Proliferation, and Federal Services. *Hearings: Report of the Interagency Review Group on Nuclear Waste Management.* 96th Cong., 1st sess., 1979.

————. Subcommittee on Nuclear Regulation. *Hearings: Nuclear Waste Management.* 95th Cong., 2d sess., 1978.

————. Subcommittee on Nuclear Regulation. *Hearings: Nuclear Waste Disposal.* 96th Cong., 2d sess., 1980.

————. Subcommittee on Nuclear Regulation. *Hearings: National Nuclear Waste Policy Act of 1981.* 97th Cong., 1st sess., 1981.

————. Subcommittee on Science, Technology, and Space. *Hearings: Nuclear Waste Disposal.* 95th Cong., 2d sess., 1978.

U.S. Department of Energy. *Backgrounder: Studies of Alternative Methods of Radioactive Waste Disposal.* Washington, D.C.: U.S. DOE Office of Civilian Radioactive Waste Management, 1986.

——. *Bulletin.* Washington, D.C.: U.S. DOE Office of Civilian Radioactive Waste Management, September & December 1986.

——. *Draft Mission Plan Amendment.* Washington, D.C.: U.S. DOE Office of Civilian Radioactive Waste Management, 1987.

——. *Draft Report to the Secretary of Energy on the Conclusions and Recommendations of the Advisory Panel on Alternative Means of Financing and Managing (AMFM) Radioactive Waste Management Facilities.* Washington, D.C.: Office of Civilian Radioactive Waste Management, 1984.

——. *Environmental Assessment: Davis Canyon, Utah.* Washington, D.C.: U.S. DOE Office of Civilian Radioactive Waste Management, 1986.

——. *Environmental Assessment: Deaf Smith, Texas.* Washington, D.C.: U.S. DOE Office of Civilian Radioactive Waste Management, 1986.

——. *Environmental Assessment: Hanford, Washington.* Washington, D.C.: U.S. DOE Office of Civilian Radioactive Waste Management, 1986.

——. *Environmental Assessment Overview: Richton Dome Site, Mississippi.* Washington, D.C.: U.S. DOE Office of Civilian Radioactive Waste Management, 1986.

——. *Environmental and Socioeconomic Considerations of Locating a Nuclear Waste Repository near Canyonlands National Park, Utah.* Washington, D.C.: U.S. DOE Office of Civilian Radioactive Waste Management, 1984.

——. *Information Services Directory.* Washington, D.C.: U.S. DOE Office of Civilian Radioactive Waste Management, 1986.

——. *Mission Plan for the Civilian Radioactive Waste Management Program.* Washington, D.C.: U.S. DOE Office of Civilian Radioactive Waste Management, 1:1985.

——. *Mission Plan Amendment.* Washington, D.C.: U.S. DOE Office of Civilian Radioactive Waste Management, 1987.

——. Press Release #R-86-065, 15 April 1986. Washington, D.C.: U.S. DOE Office of the Press Secretary.

——. *Transportation Institutional Plan.* Washington, D.C.: U.S. DOE Office of Civilian Radioactive Waste Management, 1986.

U.S. Environmental Protection Agency. *Siting of Hazardous Waste Management Facilities and Public Opposition.* Washington, D.C: EPA Office of Water and Waste Management, 1979.

U.S. General Accounting Office. *The Energy Department's Office of Environment Does Not Have a Large Role in Decision-Making*. EMD-80-50. Washington, D.C.: U.S. GAO Office of the Comptroller General, 1980.

―――. Letter to Rep. Richard Ottinger, chairman, U.S. House Subcommittee on Energy Conservation and Power. "Re: DOE Needs to Evaluate Fully the Waste Management Effects of Extending the Useful Life of Nuclear Fuel." 27 January 1984.

―――. *The Nation's Nuclear Waste: Proposals for Organization and Siting*. EMD-79-77. Washington, D.C.: U.S. GAO Office of the Comptroller General, 1979.

―――. *Nuclear Health and Safety: Oversight at DOE's Nuclear Facilities Can Be Strengthened*. GAO/RCED-88-137. Washington, D.C.: U.S. GAO, 1988.

―――. *Nuclear Health and Safety: Stronger Oversight of Asbestos Control Needed at Hanford Tank Farms*. GAO/RCED-88-150. Washington, D.C.: U.S. GAO, 1988.

―――. *Nuclear Waste: DOE Has Terminated Research Evaluating Crystalline Rock for a Repository*. GAO/RCED-89-148. Washington, D.C.: U.S. GAO, 1989.

―――. *Nuclear Waste: DOE's Handling of Hanford Iodine Information*. GAO/RCED-88-158. Washington, D.C.: U.S. GAO, 1988.

―――. *Status of the DOE's Implementation of the NWPA of 1982*. GAO/RCED-85-116. Washington D.C.: U.S. GAO, 1985.

U.S. Nuclear Regulatory Commission. *Public Information Circular for Shipments of Irradiated Reactor Fuel*. NUREG-0725. Washington, D.C.: U.S. NRC, 1984.

―――. *News Releases*. NUREG/BR-0032, vol.9, no. 28. Washington, D.C.: U.S. NRC, 11 July 1989.

U.S. Office of Technology Assessment. *Managing Commercial High-Level Radioactive Waste*. Washington D.C.: U.S. Congress, 1982.

―――. *Nuclear Power in an Age of Uncertainty*. Washington, D.C.: U.S. Congress, 1984.

Van der Pligt, J., J. Eiser, and R. Spears. "Public Attitudes to Nuclear Energy." *Energy Policy* 12, no. 3 (1984): 302–05.

Walker, C., L. Gould, and E. Woodhouse, eds. *Too Hot to Handle?: Social and Policy Issues in the Management of Radioactive Waste*. New Haven: Yale University Press, 1983.

Watkins, J., Secretary of Energy. Remarks. Washington, D.C.: U.S. DOE, 1989 Press release R-89-068.

Weinberg, A. "Social Institutions and Nuclear Energy." *Science* 177, no. 4043 (1972): 34.

Wolpert, J. "Departures from the Usual Environment in Locational Analysis." *Annals of the Association of American Geographers* 60, no. 2 (1970): 222.

Ziegler, D., J. Johnson Jr., and S. Brunn. *Technological Hazards.* Washington, D.C.: Association of American Geographers, 1983.

INDEX

225

Pitt Series in Policy and Institutional Studies
Bert A. Rockman, Editor